HOME PLUMBING REPAIRS

by Larry Elrod

THEODORE AUDEL & CO.
a division of

HOWARD W. SAMS & CO., INC.
4300 West 62nd Street
Indianapolis, Indiana 46268

FIRST EDITION

FIRST PRINTING—1974

International Standard Book Number: 0-672-23802-0

Contents

SECTION 1

Introduction

Many states and cities require licensed plumbers to do plumbing installation and repairs in commercial buildings and public places. Most building or plumbing codes also require plumbers to install new domestic plumbing when houses are built. However, virtually all homeowners *can* do their own plumbing repairs and need not be a licensed plumber.

With a little effort at learning and some practical common sense you can repair your plumbing. It isn't all that difficult. The neighborhood kid who has worked on cars for 10 years (since he was 15) probably was self-taught. If he does it regularly, he probably is pretty fair at his repair work. Regardless of how good he gets he puts up with the "shade-tree mechanic" title because he knows how much money he is saving. You too can piddle around with plumbing at your house and get pretty good at it. Who knows, you too may even earn that seemingly contemptuous title from your neighbors of "shade-tree plumber," but just think of the money you will be saving. The real and honest test of your ability and recognition as a do-it-yourself plumber comes when your friends and neighbors start coming to you for *free* advice.

Plumbing repairs are never inexpensive: neither are parts and supplies on small repair jobs that you do yourself. However, by doing it yourself, you can literally save a bundle of cash. The information and tips in this book are not meant to make you a journeyman plumber, but they should save you money and give you a limited working knowledge to help yourself.

Unless you have previous plumbing experience, repair work *beyond* the tips presented here should be approached with caution and with the phone number of the local plumber written on the wall by the phone.

TOOLS

Since you want to do the job right, a few specific tools will enable you to do so. Many of them you may already have. Good screwdrivers of the slot- and phillips-head type are required (Fig. 1). You should also have at least two wrenches. One is a 10″ pipe wrench (Stillson) and an adjustable end wrench as shown in Fig. 2. The pipe wrench has teeth in the jaws and this is the only common wrench you can buy that will hold or turn pipe. It will often be used as a "backup" to hold something from turning while loosening or tightening something else. Pliers are useful everywhere. Standard pliers and the *Channellock®* style especially are good. The *Chan-*

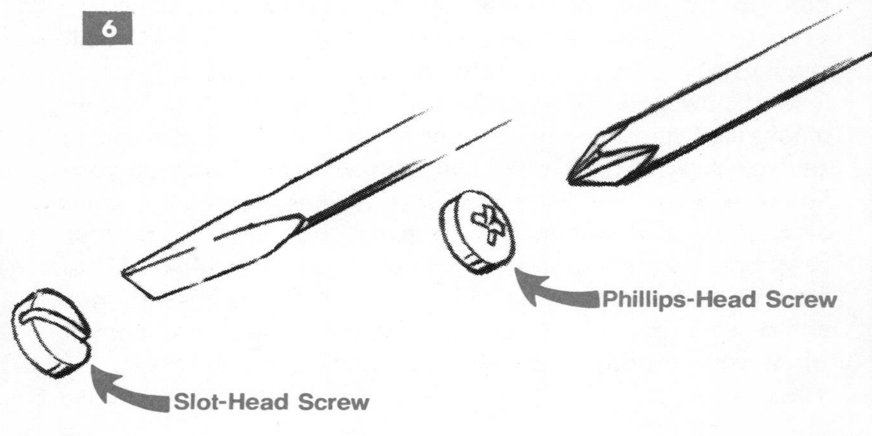

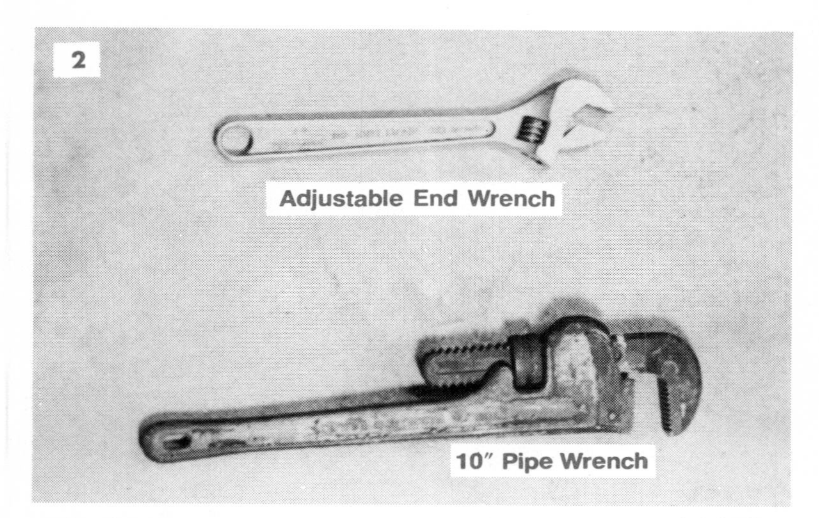

nellocks® are often used as a pipe wrench when heavy strains are not required to turn pipes (Fig. 3). *Snakes* are used to cut or bore through obstructions in drain lines, pipes, toilets, etc. A short snake, or *closet auger,* is fed to unclog most simple obstructions in household plumbing. The other kind of snake is a tightly wound and flexible instrument that can be bought in 25- and 50-foot lengths. A longer snake is necessary if you have an obstruction in a pipe between the house and sewer or septic tank. Snake costs are not high.

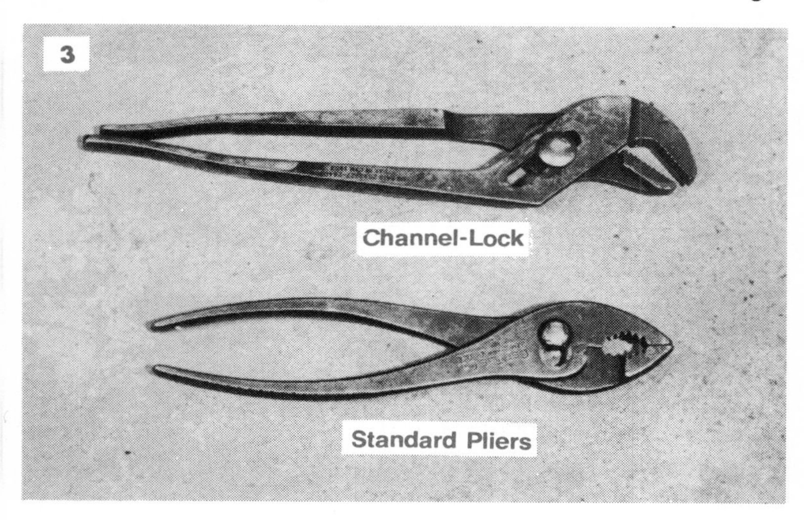

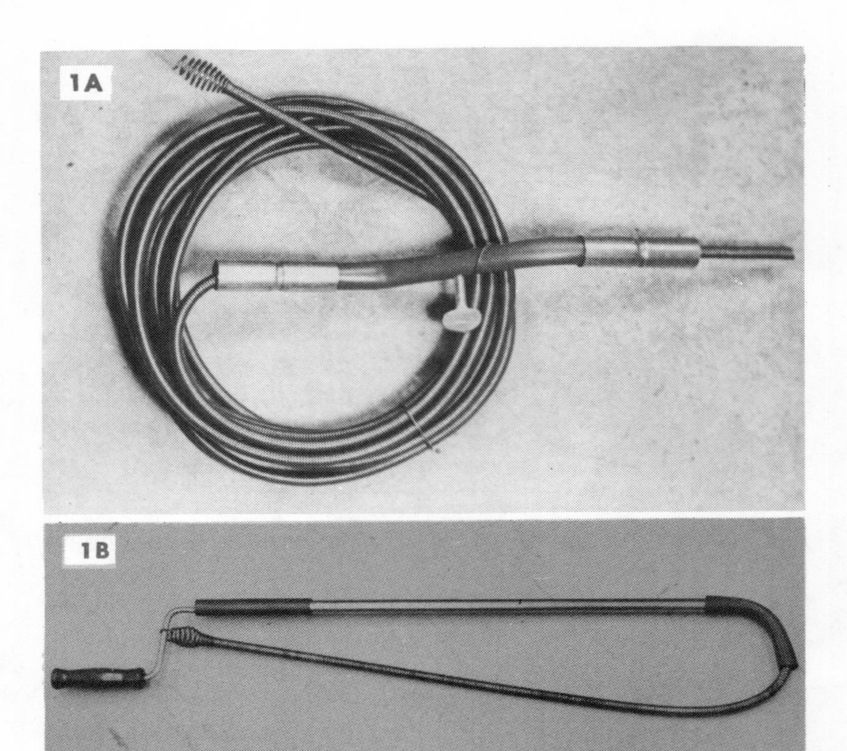

But unless you have recurring root-clogging or septic tank back-up problem you should rent exactly the snake you need (Fig. 4).

Common sense is another tool you should have. It will enable you to improvise all kind of gadgets to help you out. Some of which are wire hangers of any kind for supporting pipes and hoses strung overhead, and "coat hanger" snakes.

Everyone should have a "plumbers friend," or a plunger (Fig. 5), along with a small assortment of other tools and supplies like hammer, hacksaw, files, plastic tape, pipe dope of the *Blue Ribbon* or *John Sunshine* type, tape pipe dope, epoxies, tub caulking, solder, flux, a small torch, and emery cloth (Fig. 6). Seldom will you need a torch unless a leaky copper pipe joint needs fixing or an entire faucet needs replacing. In any case, a small torch is a handy tool to have for all kinds of heating jobs around the house.

5

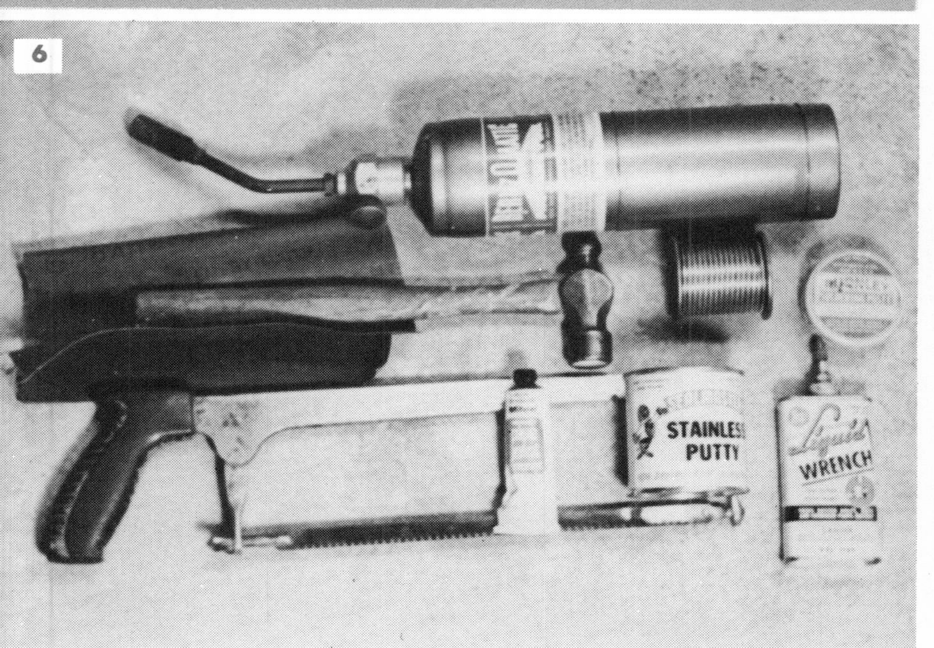
6
PUTTY
WRENCH

SECTION 2

Clogged Drains, Toilets and Waste Lines

When the water won't drain away from the lavatory, sink, or the flushed toilet fills up the bowl and runs over onto the floor, it doesn't take long to figure out that something is clogged up. Unclogging sink and lavatory drains are similar operations.

LAVATORY

The most common problems in lavatory drains is the accumulation of hair caught in the stopper or hairpins and other items in the trap. First remove the stopper by pushing down on it slightly and turning left then pulling straight up and out (Fig. 1). If the stopper will not come out, it is a different type. If so, unscrew the sleeve nut, pull out the lift arm and disengage it from the stopper (Fig. 2). Lift out the stopper. If there is hair, string, and other junk building up

1A

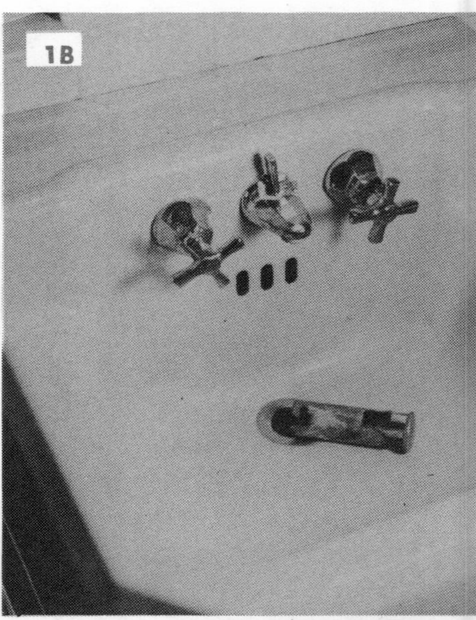
1B

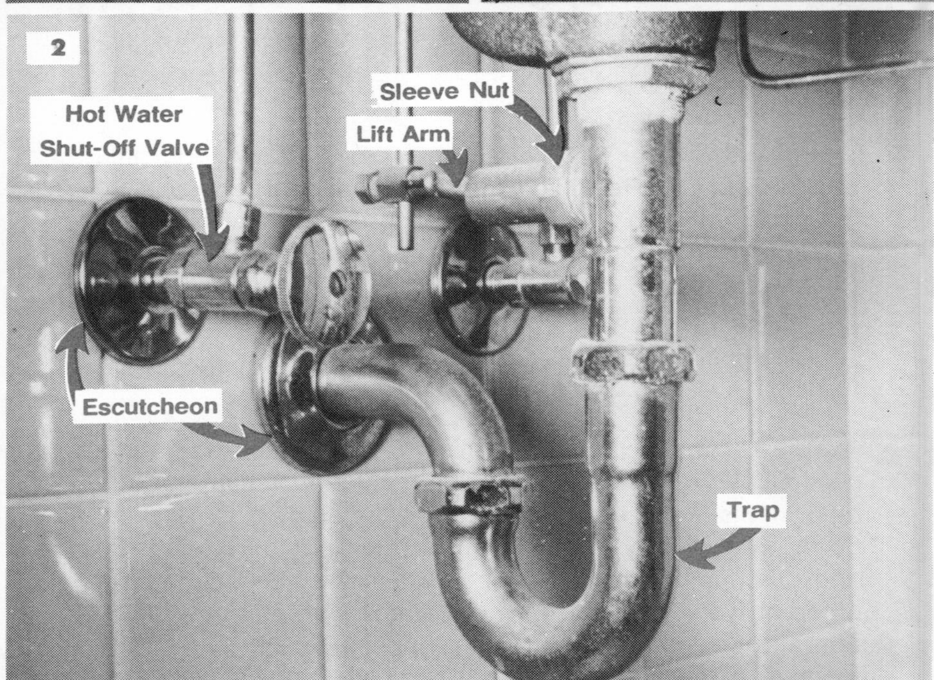
2
Hot Water
Shut-Off Valve
Sleeve Nut
Lift Arm
Escutcheon
Trap

on the stopper, clean it thoroughly. Replace the stopper, insert the lift arm, and screw on the sleeve nut (Fig. 2). Test for proper operation. If this did not clear up the problem, it could be a clogged trap; try a plunger.

Plungers only work if they can push and pull a small amount of water back and forth in a trap, pipe, or waste line. There probably already is some water in the lavatory bowl if there is a clog in the drain; if not, put at least three inches of water in the sink and place the plunger over drain and push fully up and down with sharp forceful strokes (Fig. 3). If the lavatory has an overflow cast into it, a plunger won't work.

There are several good commercial drain cleaners that can be used. Some are liquid and others are granulated. Carefully follow the manufacturer's directions for use (Fig. 4). If this frees the obstruction, be sure to run hot water down the drain to flush all remains of the drain cleaner from the trap. If this does not clear the clogged trap, run a small

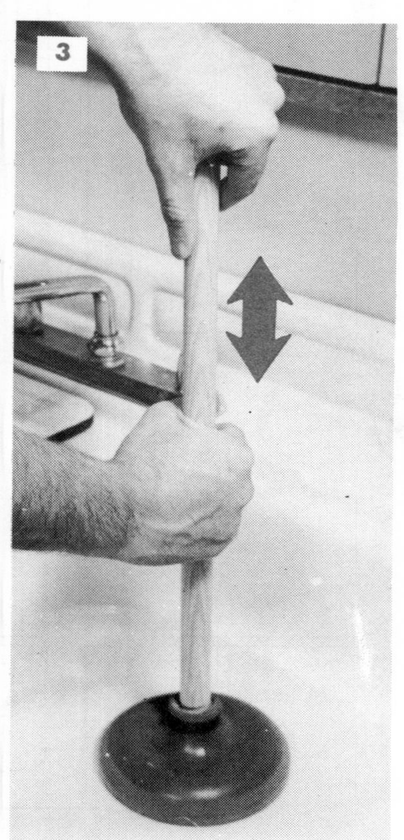

snake into it. Apply pressure into the trap with the snake from above, turn the snake vigorously so it will rotate and "screw" its way through the obstruction (Fig. 5). If the drain is cleared, flush with much hot water to wash away the broken pieces of the obstruction.

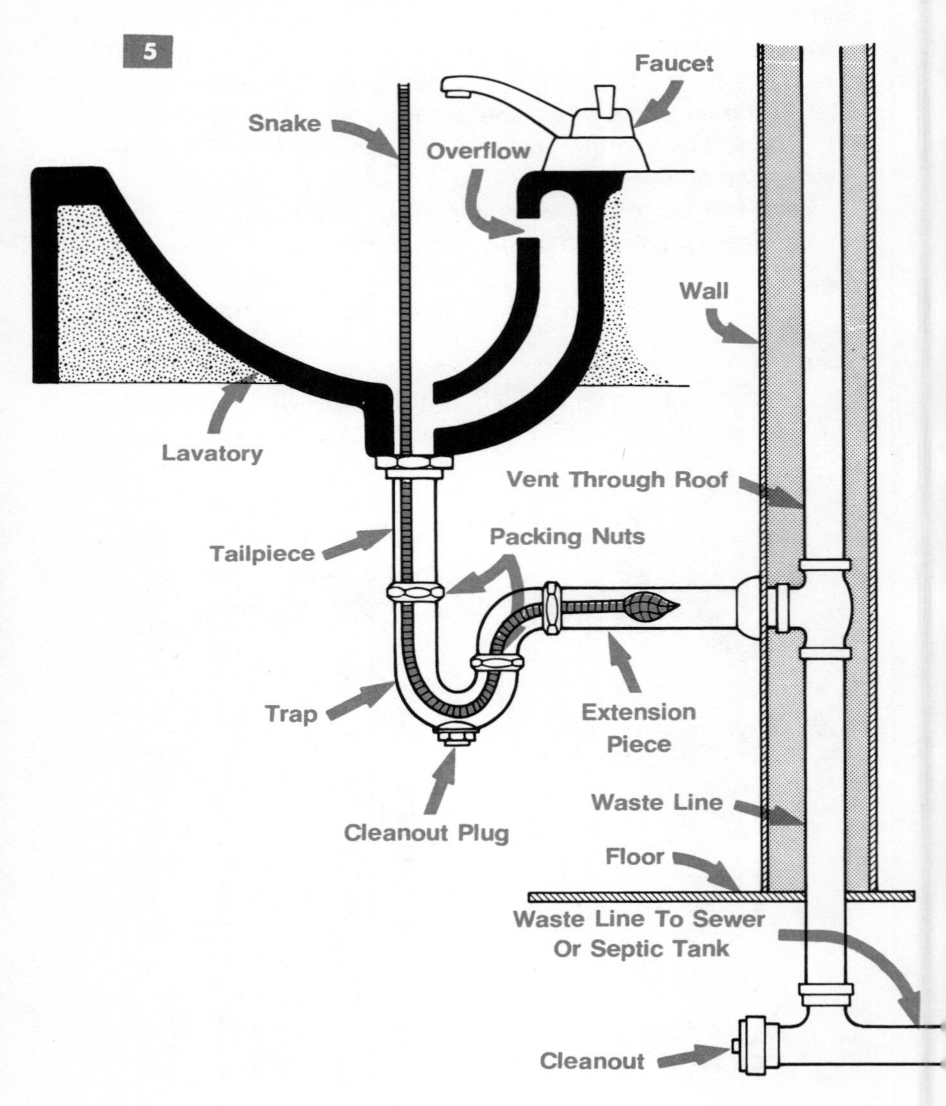

If the trap is still clogged and a snake won't go through it, the trap must be removed. *Caution.* If you have used a drain cleaner, this caustic and dangerous solution is still in the trap. It will cause severe eye burns and mild skin burns. Wear rubber gloves, glasses, and place a pan or bucket under the trap to catch any water that runs out when the trap is removed (Fig. 6). Some traps have a clean-out plug in the bottom. Remove this plug and drain the water. Run the snake up through this plug opening to remove the obstruction. Many modern house-fixture traps do not have this cleanout plug so don't be alarmed if you don't find one.

To remove the trap, loosen and slide back the packing nuts shown in Fig. 5. Pull the trap down away from the tail piece. Clean out the trap and reassemble. The packing nuts prevent leaks by putting pressure on the washers inside them. Most are 1¼″- or 1½″-diameter washers. Determine which you have and always replace them with new ones when assembling the trap. If the washers have a bevelled edge, make sure they go back just like the originals came out. Slide the trap up on the tail piece until the threads line up with packing nut. (See Fig. 5). Tighten packing nuts *A* and *B* and run water to test for leaks. Use caution when you throw out that pan of caustic water that contains the drain cleaner.

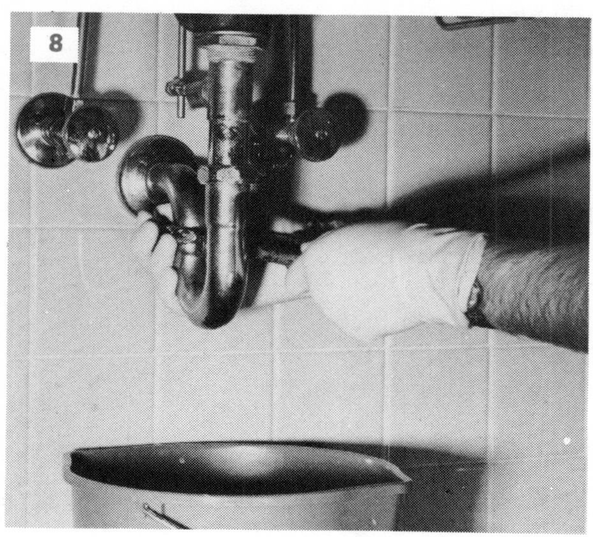

If you discover after all of this that the obstruction is beyond the wall, remove the trap and run a snake into the extension piece and down into the waste line shown in Fig. 7. Remember to twist the snake vigorously to get it to "feed" *down* and around inside the pipe. Make sure the snake does not go *up* the vent pipe. In very rare instances, a clogged vent to the roof will cause problems. Check it.

Now is the time to think "plumbing care" about your bathroom lavatory drain. To prevent clogs, keep stopper in place so large objects won't go into the trap. When combing hair in front of the mirror, stand back a step so hair will fall on floor and not in the lavatory. Make sure all family members are aware that absolutely nothing goes down

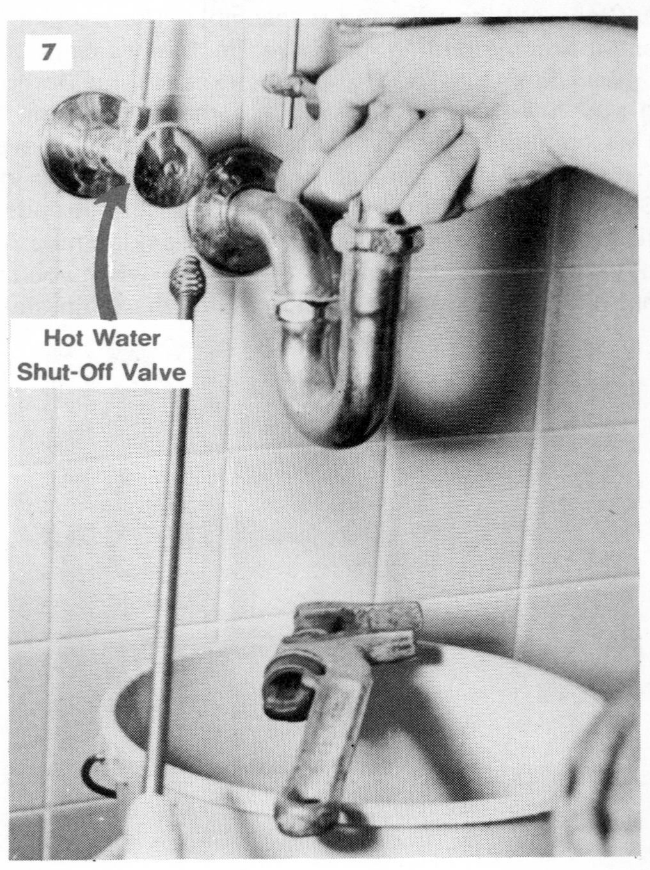

a drain except soap and water. Some people have actually poured latex paint and plaster down drains or toilets. This obviously is a serious no-no-. Over the months and years, daily use of "sand" hand-cleaning powders and soaps can build up sand in a trap.

SINKS

Sinks in the kitchen have drains that are serviced exactly like bathroom lavatories and have many of the same problems. There are many exceptions. Sinks handle grease and food and lavatories do not. Sink drain troubles are usually related to grease and food congealing on the walls of cold pipes. This is a very gradual build-up and because of this build-up, grease clogs can be very hard and occupy several feet of pipe.

Try all of the trap cleaning procedures outlined above for lavatories. Then run a snake into the extension piece and down the wall pipe into the waste line as far as possible (Fig. 5). If the sink has a garbage disposal, it must be disconnected from the drain before any clog-removing procedures are started. Do not put drain cleaner in a garbage disposal unit. If the trap or waste lines are exposed to areas that are cold or not heated, try applying heat from a small space heater or high-wattage electric tape (Fig. 8). If the

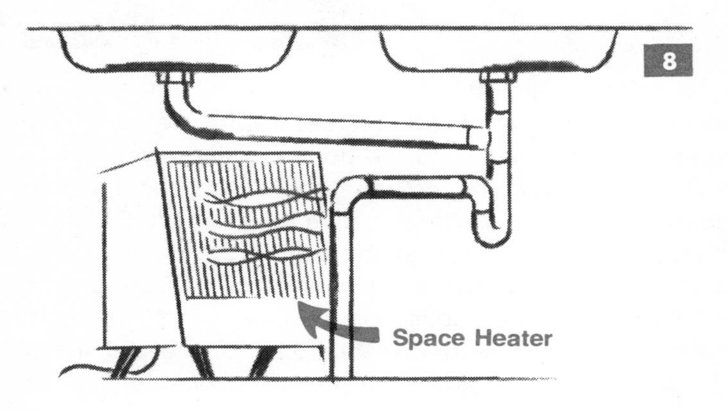

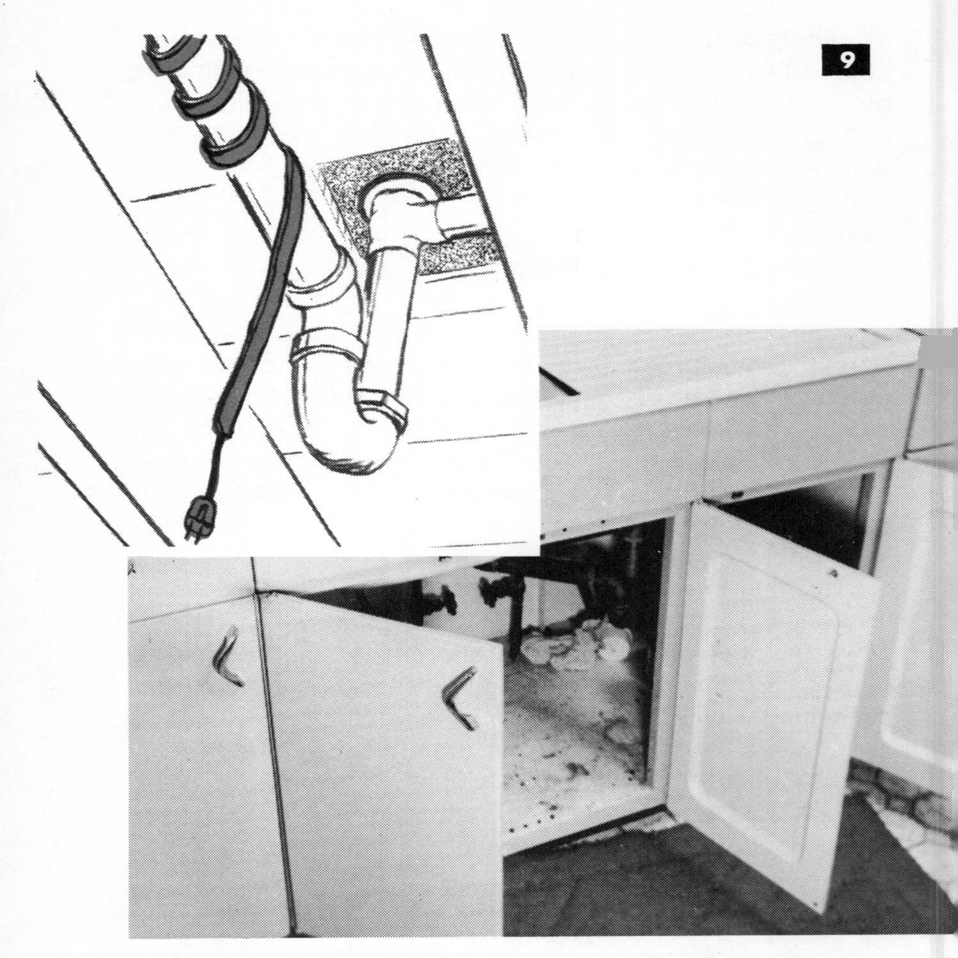

lines or trap cannot be heated properly, run in the snake using the technique mentioned before. Problems of clearing clogs in waste lines from the house to the sewers or septic tank are covered in another book. Once the blockage is cleared it is time to think "plumbing care." Examine the drain pipes and trap under the sink. If the pipes or the trap is in a nonheated space such as a crawl space or other compartment they should be wrapped with any kind of material that will provide insulation. Commercial insulation can be bought for less than the price of one visit of the sewer-line cleaning man. If waste lines in a crawl space are near enough to the ground, they should be covered with 6″ to 12″ of dirt. This isn't enough to prevent freezing outside but is usually sufficient insulation in a crawl space under the kitchen floor. Other methods of preventing grease from setting up in lines and traps is to keep the sink cabinet door open to kitchen heat during extremely cold weather (Fig. 9) and wrap waste lines with heat tape. Routine use of a drain cleaner every month or two will help to soften the build-up of grease. The *most* effective method for keeping sink drains clear is to keep the grease out of it. It is not possible to keep 100% of it out but most of it can and should be put in the garbage can. Wipe greasy skillets, pots, pans and dishes with a paper towel. Soak up that grease and throw it away. Every greasy skillet put in the sink dish water shortens the time until the next waste-line stoppage.

TOILETS

If the toilet bowl fills up with water and doesn't flush or only seeps down and out slowly, use a plunger to try to dislodge a possible obstruction in the trap (Fig. 10). If this

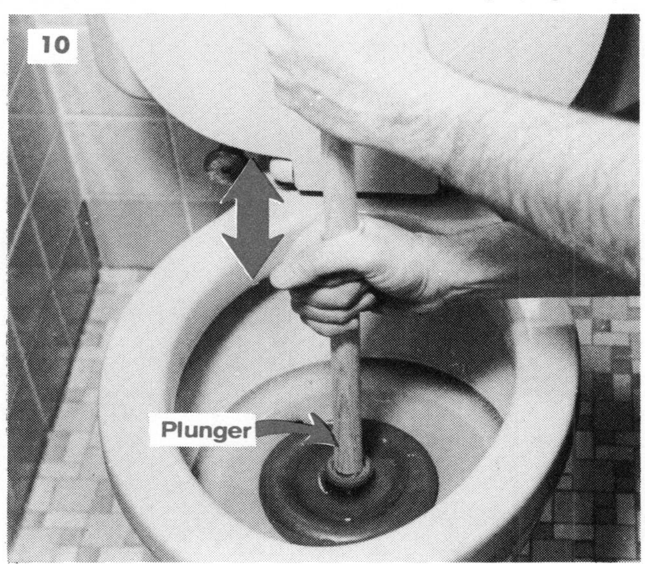

fails use a snake, or even a coat hanger in emergencies, and bore out the obstruction (Fig. 11). If it is cleared, all material will be readily flushed away. If the toilet trap was not clogged, then find the cleanout nearest the toilet waste line and use a snake. This procedure is discussed in a later section on waste lines. If your waste line is clear to the septic tank, then you have septic tank problems which are covered in another book of this series. If the line is clear and you are on a city sewer, call the city sanitation department. There may be high water backed up into the sewer or other problems.

Remember, when you remove the clean-out plug or cap, all of the water and waste material standing in the toilet and its waste line above you will come splashing down in your face. Get a bucket!

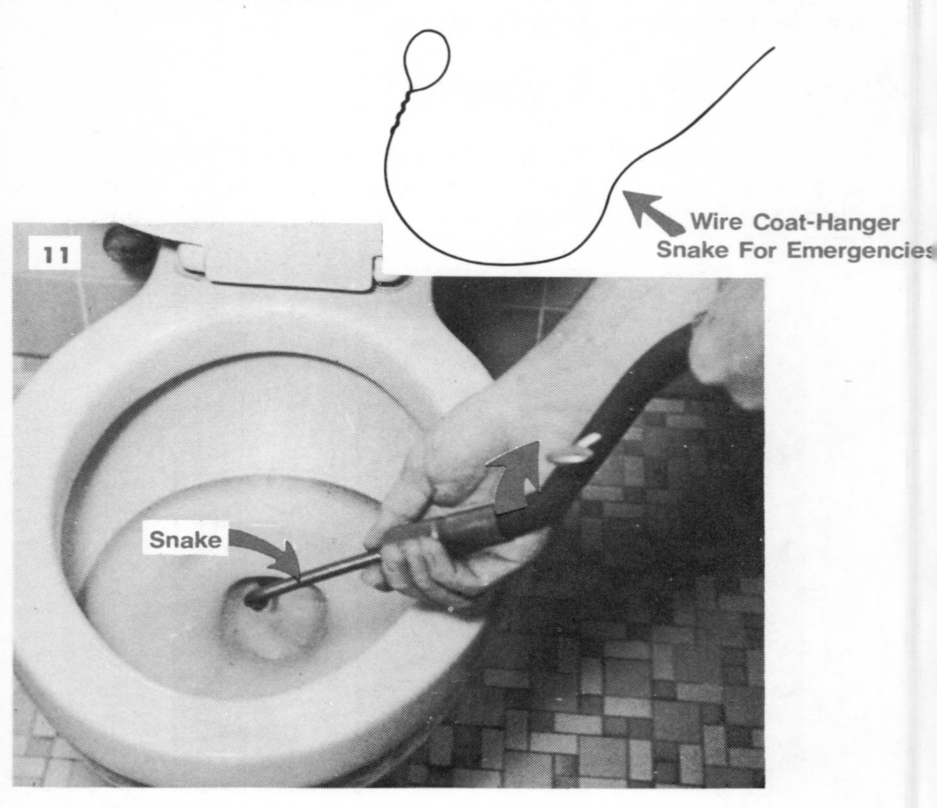

Again you must become conditioned to think "plumbing care." If you have city sewers and are connected to them, your concern over what goes into the toilet need not be as great. But if you are on a septic tank you will need to really get acquainted with the section in the companion book on septic tanks.

Toilets are designed to remove human waste and the associated paper products. They also provide a water seal in the trap to prevent foul sewer gas from coming into the house. Because of the zig-zag shape of this trap (Fig. 12), a lot of bulk things being flushed at one time may clog the trap.

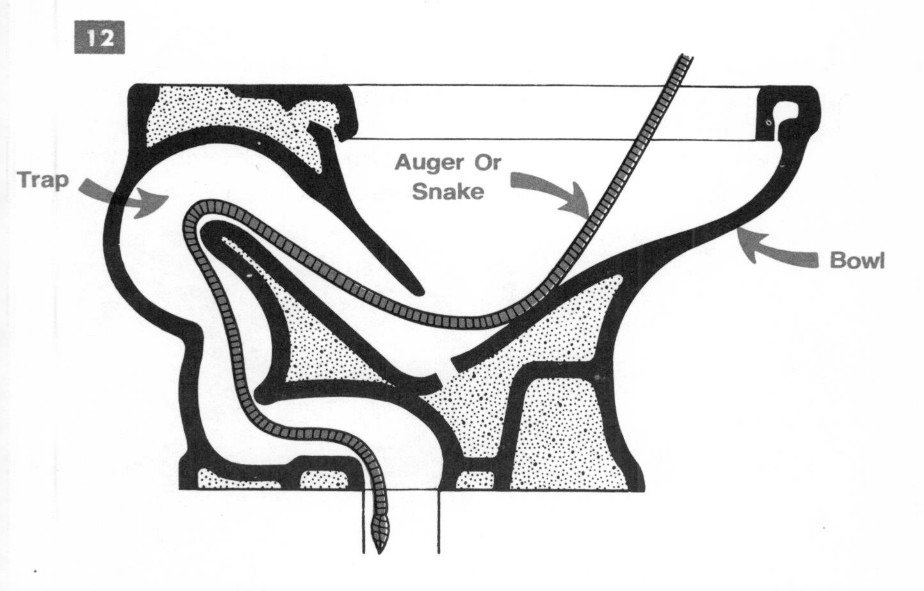

Among these things are too much toilet paper. Children and some cleanliness fanatics use far too much toilet paper and hence some toilets won't flush because of a choking wad of paper. Plastic toys, pencils, hair brushes, watches, shampoo bottles, and any number of things can provide just enough obstruction if dropped into the bowl unnoticed. When flushed with even a limited amount of paper and human waste, you have a toilet bowl running over with water. Another pair of notorious toilet stoppers are sanitary napkins and tampons. These items should not be flushed down because of their bulk. Whether wrapped in toilet paper or not, they can produce a clog and often do. They don't do a septic tank much good either if they make it that far. In recent years a few sanitary napkin makers have come out with "flushable" napkins. These have more decomposable paper content than the other type and provide less flushable bulk. Don't allow it in your sanitation system regardless what the advertisements say. Keep that stuff out of the toilet and you will have no fear.

TUBS

Tub drains and traps are generally the most trouble-free. However, they do collect hair and objects like hair pins, plastic toys, etc. like any other drain or trap. If you are going to work on or in a tub be sure to line the bottom with newspapers or an old rug (Fig. 13) to keep from scratching the bottom. Better yet remove your shoes if you get into a tub.

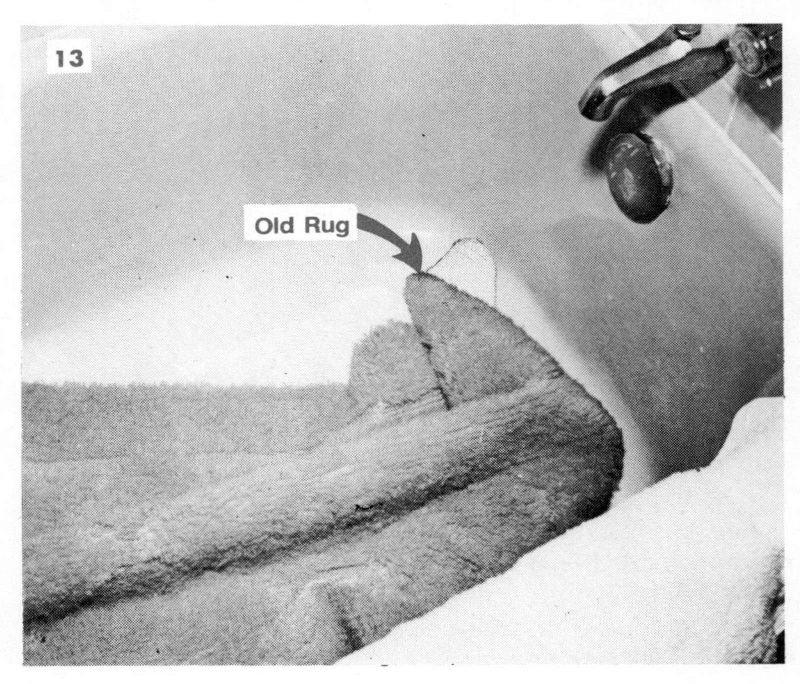

Tub drain clogs can be cleared in the same manner listed under lavatories. Use a good drain cleaner and then a snake if that fails. If necessary disassemble the trap, which will usually be found below the floor, by loosening the locknuts and pulling the trap down and away (Fig. 14). Some tubs have a strainer that will need to be removed first. Others will have a pop-up lavatory type stopper.

Access to tub drains and traps are in the basement (Fig. 14), through an access panel behind the tub located in another room or closet (Fig. 15), or in a crawl space under the house.

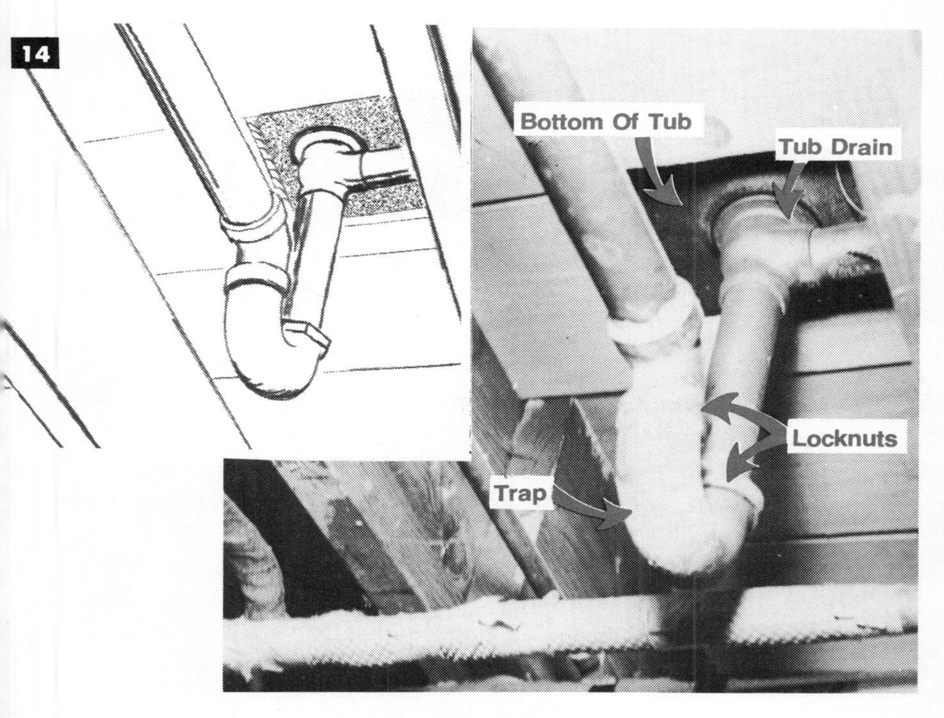

FLOOR DRAINS

Floor drains that run directly to a sewer always have a trap under the floor with standing water in it to prevent sewer gas from coming into the house. Floor drains that run to a sump usually do not have a trap. If the floor drain stops

up regardless of where it goes, remove the screen (Fig. 16) and try a good drain cleaner providing you can get any amount of water in it at all. If this fails or you can't get any water into it, use a good snake. Once the snake clears the obstruction, flush with lots of water.

Floor drains are not often used. Those with a trap often have the water evaporate to a low level or even dry. This then lets sewer gas in the house. To ensure a proper seal, pour a bucket of water in the drain regularly.

WASTE LINES

The pipes running from your house to the sewer or septic system are subject to a couple of important problems—tree roots and grease. Access to the large pipes (usually 2 to 4 inches) is through plugs in the ends of these pipes and are

called cleanouts (Fig. 17). The most common is a brass plug that can be removed with a good-sized wrench. To chew up or round off the lug on a clean-out plug simply means cutting it out with a hammer and chisel. That is by no means a fun job.

To remove a stubborn cleanout plug, follow these instructions. Apply penetrating oil to the threads of the plug (Fig. 18) and tap the plug on top a couple of times with a hammer to

"loosen up" any corrosion. Let this set for a few minutes and apply more oil and tap the plug again. Put a wrench on the plug and put a heavy strain on it counterclockwise (Fig. 19). If it does not move, put a heavy strain on the wrench while lightly tapping the soil-pipe hub or side of the cleanout with a hammer.

Once the plug is out, insert a snake long enough to reach the sewer or septic tank. Again, run the snake in with a vigorous circular twisting motion to "bore through" an obstruction (Fig. 20). If you find it impossible to get through the obstruction, pull back the snake carefully a couple of times and inspect the end (Fig. 21). A simple coil-ended snake will bore through a paper obstruction and not much else. If you find small hairlike tree roots, go to the tool-rental store and get a motordriven snake with a cutting head on the end. Better yet, call a professional roto-rooter or sewer-line cleaning company because power snakes can be dangerous if you don't know what you are doing. For a sink waste line that is clogged with grease, a power snake will be needed.

Once you have cleared the obstruction, insert a garden hose and flush with a large amount of water (Fig. 22). Put a little vaseline on the cleanout plug threads and put it back

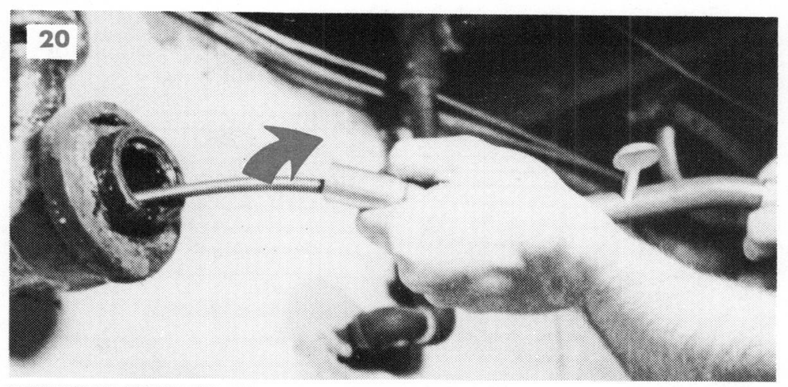
20

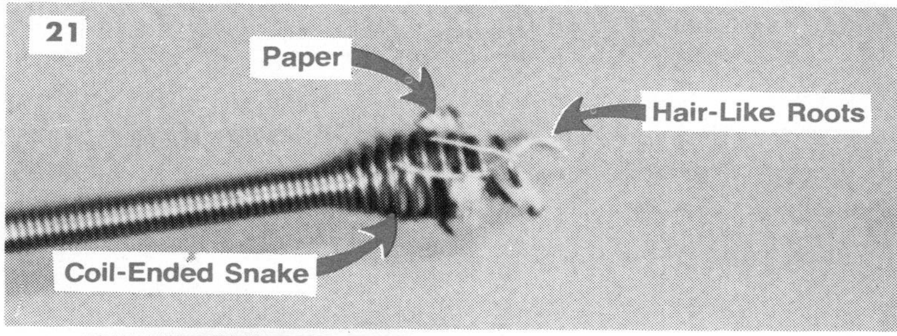
21
Paper
Hair-Like Roots
Coil-Ended Snake

22
Garden Hose

just snug. It does not need to be extremely tight. If you found tree roots in the line, you can plan on doing this job routinely. How often depends on how many roots have found their way into your line. Roots get into these lines through poor joints between pipes, cracks or breaks caused by ground shifts, and other reasons. Ultimately, to stop it, a new pipe will have to be installed.

Again think "plumbing care." When you put that cleanout plug back in, check the lug. If it is "dog-eared" or chewed up, put in a new one. Take it to a plumbing supply store and get an exact replacement. Think in terms of taking it out the next time.

If you moved into a house that is a few years old, the trees have already been planted. If you have a septic system, check where it and the trees are. If there is a particularly severe root problem, you may need to remove some trees. By all means, if you are in a new house, don't plant trees over your septic system.

SECTION 3

Sink, Lavatory and Tub Problems

The problems encountered with sinks, lavatories, and tubs are all very similar. All have faucets of some sort and once you are convinced that they work about the same and understand them, they all can be fixed or replaced. All sinks, lavatories, and tubs have some means of keeping the water in the bowl or tub by means of a drain-stopping device. The most common of these is a mechanical stopper with movable linkage. Below the base of the bowl all sinks, lavatories, and tubs have drainpipes with slip nuts and washers. Once you understand the simple principles of drain fittings, all problems are solved virtually alike.

FAUCET PROBLEMS

Most kitchen sinks are of the double basin type and have a faucet with a swing neck that will reach both sides (Fig. 1). The bathroom lavatory faucets are usually of the type shown in Fig. 2. An entire kitchen sink installation can be seen in Fig. 3. A typical lavatory installation can be seen in Fig. 4.

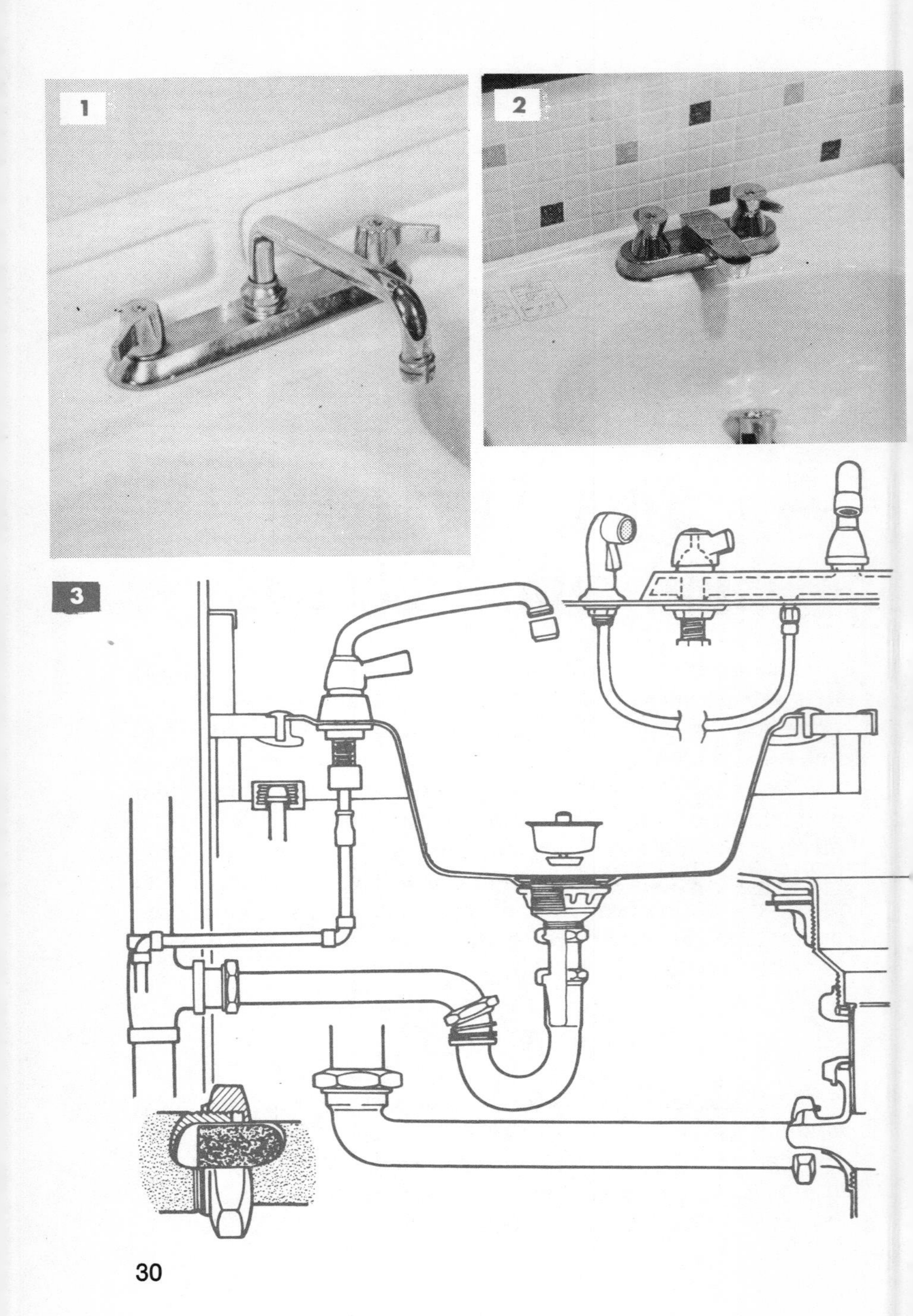
1
2
3

4

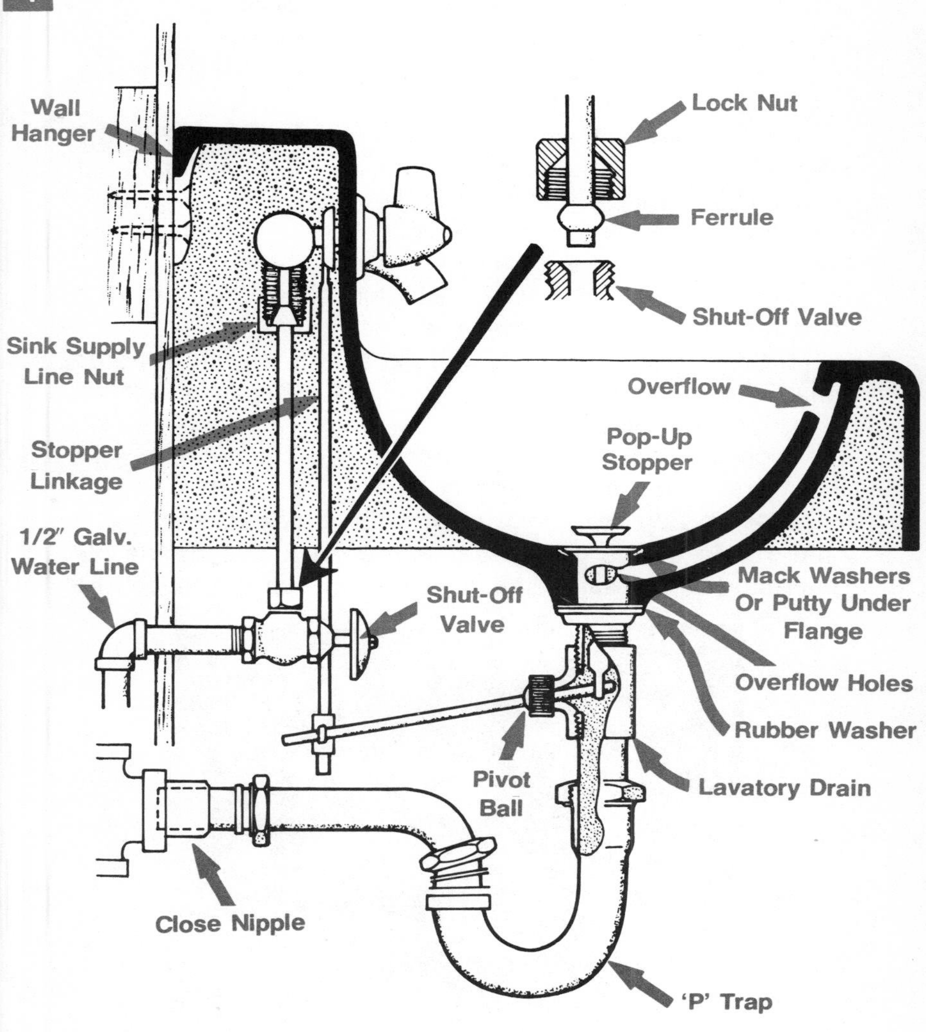
Wall Hanger
Lock Nut
Ferrule
Shut-Off Valve
Sink Supply Line Nut
Overflow
Stopper Linkage
Pop-Up Stopper
1/2" Galv. Water Line
Mack Washers Or Putty Under Flange
Shut-Off Valve
Overflow Holes
Rubber Washer
Pivot Ball
Lavatory Drain
Close Nipple
'P' Trap

Compression Faucets

The most common problem with any two-handle compression faucet is that it begins to drip from the spout or leak around the stems. These problems are also encountered in the single-lever type faucet, only less frequently.

If a leak develops and drips from the spout, shut off the water to that faucet and do the following:

Step 1. Remove the screw that holds the handle to the stem (Fig. 5). Sometimes to get at the screw, a small pop-out cap must be removed in the top of the handle.

Step 2. Pull handle straight off of stem. Careful prying with a screwdriver may be needed.

Step 3. If faucet has an escutcheon, remove it. The packing nut (Fig. 5) will now be exposed.

Step 4. Remove the packing nut with a wrench or in the case of a recessed tub faucet a $1\frac{13}{16}$ or $1\frac{29}{32}$ socket wrench (Fig. 6) must be used.

Step 5. Twist back out and remove the stem (Fig. 7). This has the seat washer on it (Fig. 7).

Step 6. Replace the washer with a new one (Figs. 8 and 9). Make sure the washer is the exact size as the old one and has the same shape. Depending on the faucet, be

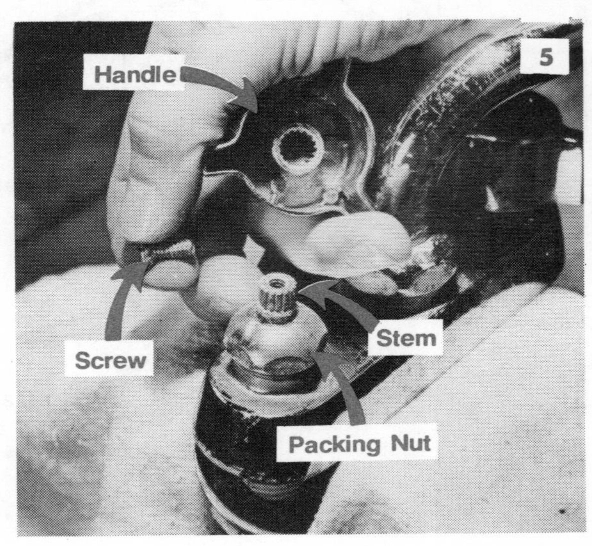

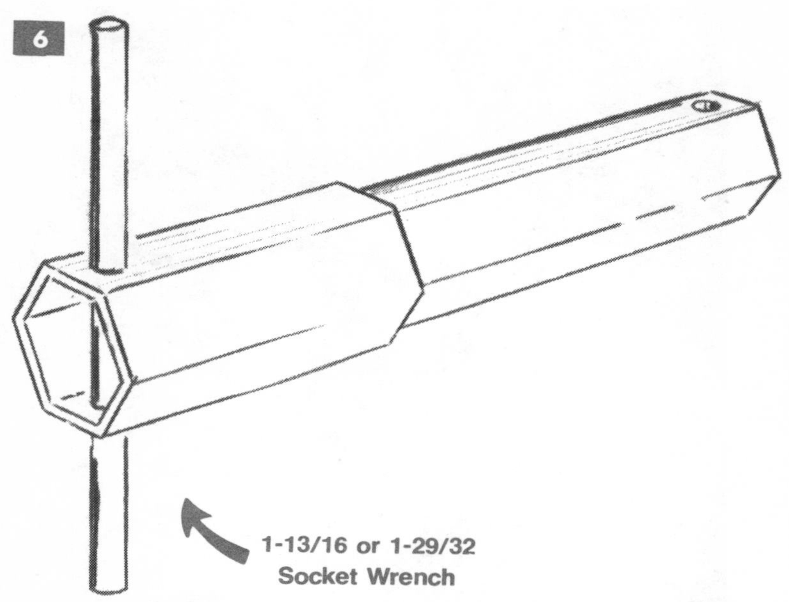
6
1-13/16 or 1-29/32
Socket Wrench

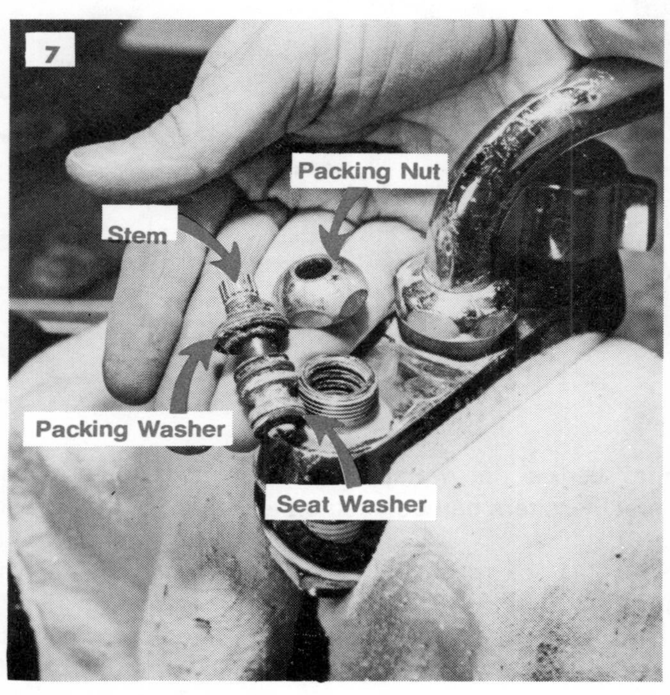
7
Packing Nut
Stem
Packing Washer
Seat Washer

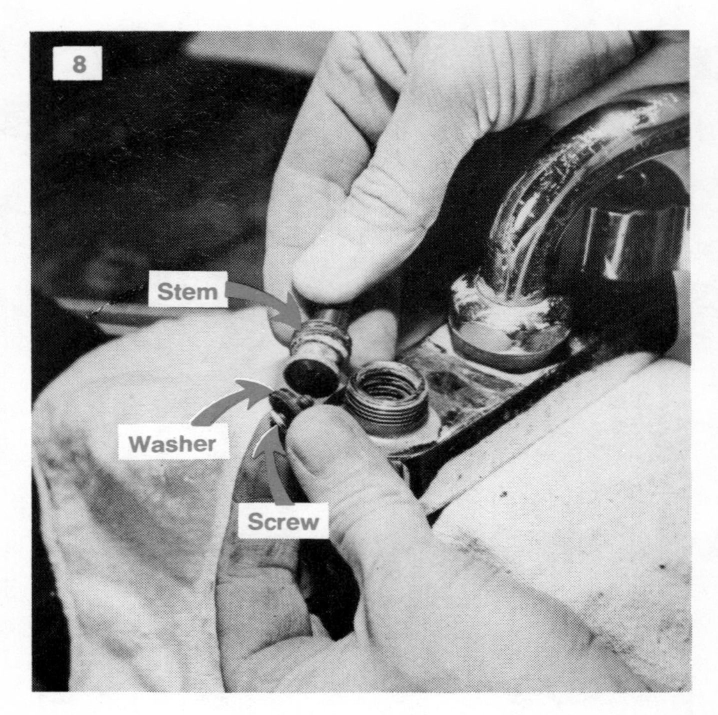

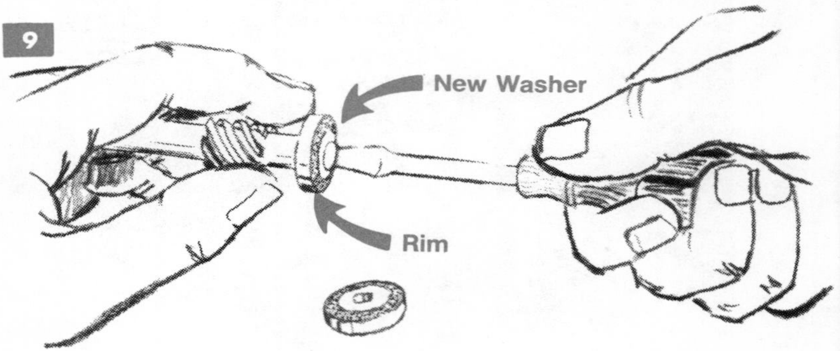

sure to use a hot water washer or cold water washer. Universal washers can be used either as hot or cold washers (Fig. 10).

Step 7. Look down into the faucet where the stem came from. Look at the seat to see if it is smooth (Fig. 11). Run your fingernail around the seat to check for deposits or cuts in the surface. If the seat is damaged, carefully resurface

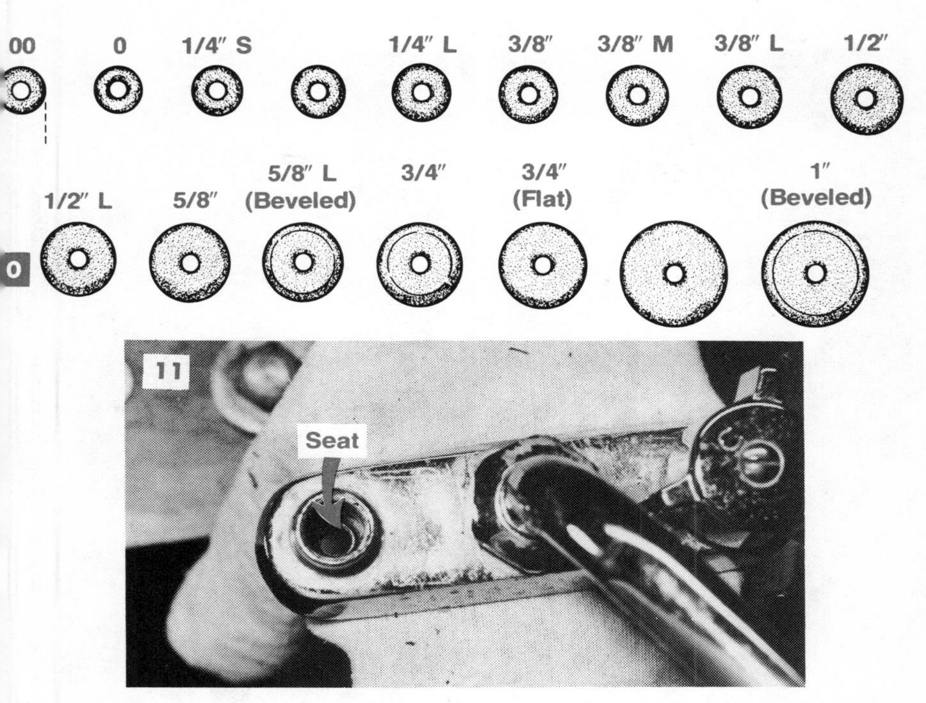

it. Use a resurfacing tool as shown in Figs. 12 and 13. Set the cutting head on the seat and slowly turn the handle. Be careful, don't cut too deeply.

Step 8. If seat can't be reseated replace it. First remove seat with a seat wrench (Fig. 14). Take the seat to a plumbing supply house and get an exact replacement.

Step 9. Return the stem with a new washer on it to the faucet. If the stem is damged or corroded, an entire replacement stem can be purchased (Fig. 15). Take the old one with you to a plumbing supply house and get an exact replacement. Be sure to know the make of your faucet.

Step 10. Replace the packing washer with a new one if possible. If not, use the old packing. If this leaks, use a stranded graphite wicking (Fig. 16) or string around the stem spindle.

Step 11. Turn the packing nut down snug against packing to smash or compact it around the stem to form a watertight seal.

12
Seat Cutter

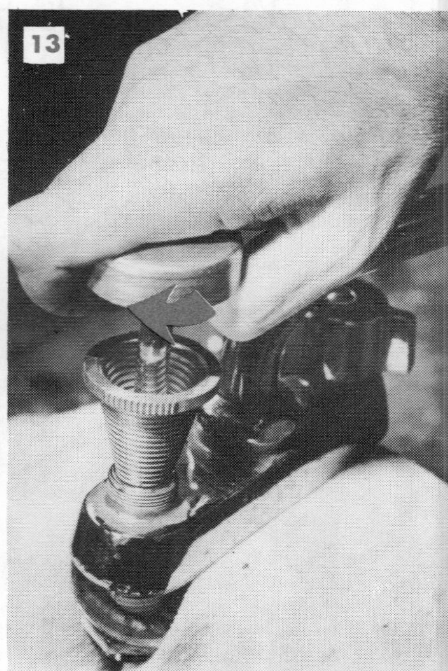
13

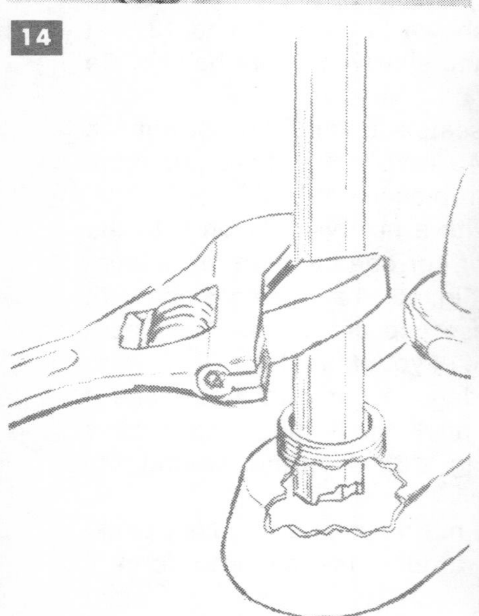
14

14
Seat Wrench

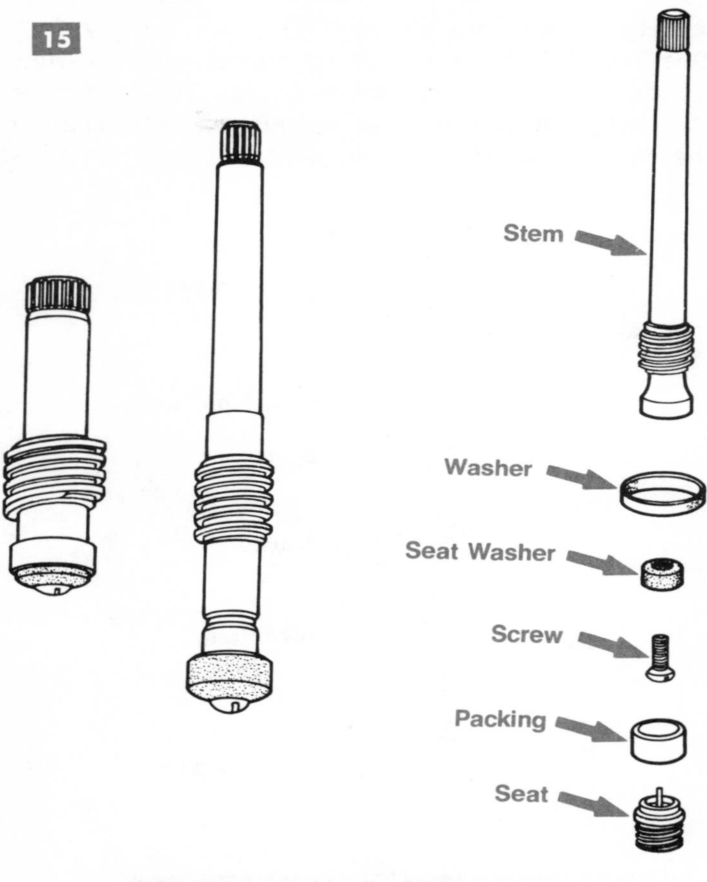
15
Stem
Washer
Seat Washer
Screw
Packing
Seat

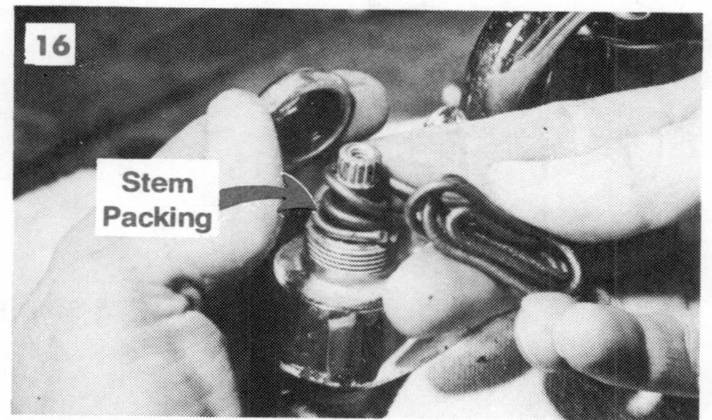
16
Stem
Packing

This is the general repair procedure for most compression faucets whether they are used on a tub, lavatory, or a sink. See Figs. 17, 18, 19, and 20 for typical exploded-view installations. Carefully study these figures and recognize common parts that appear in all faucets. Faucets different from the ones shown will have the same or similar functioning parts.

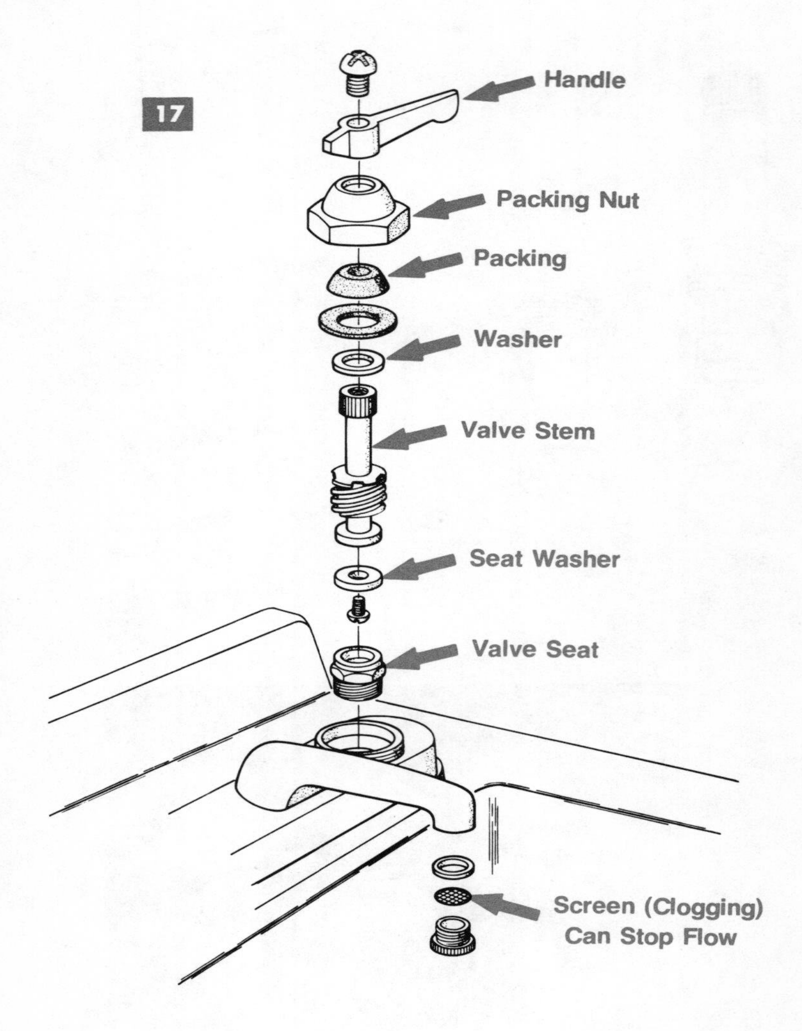

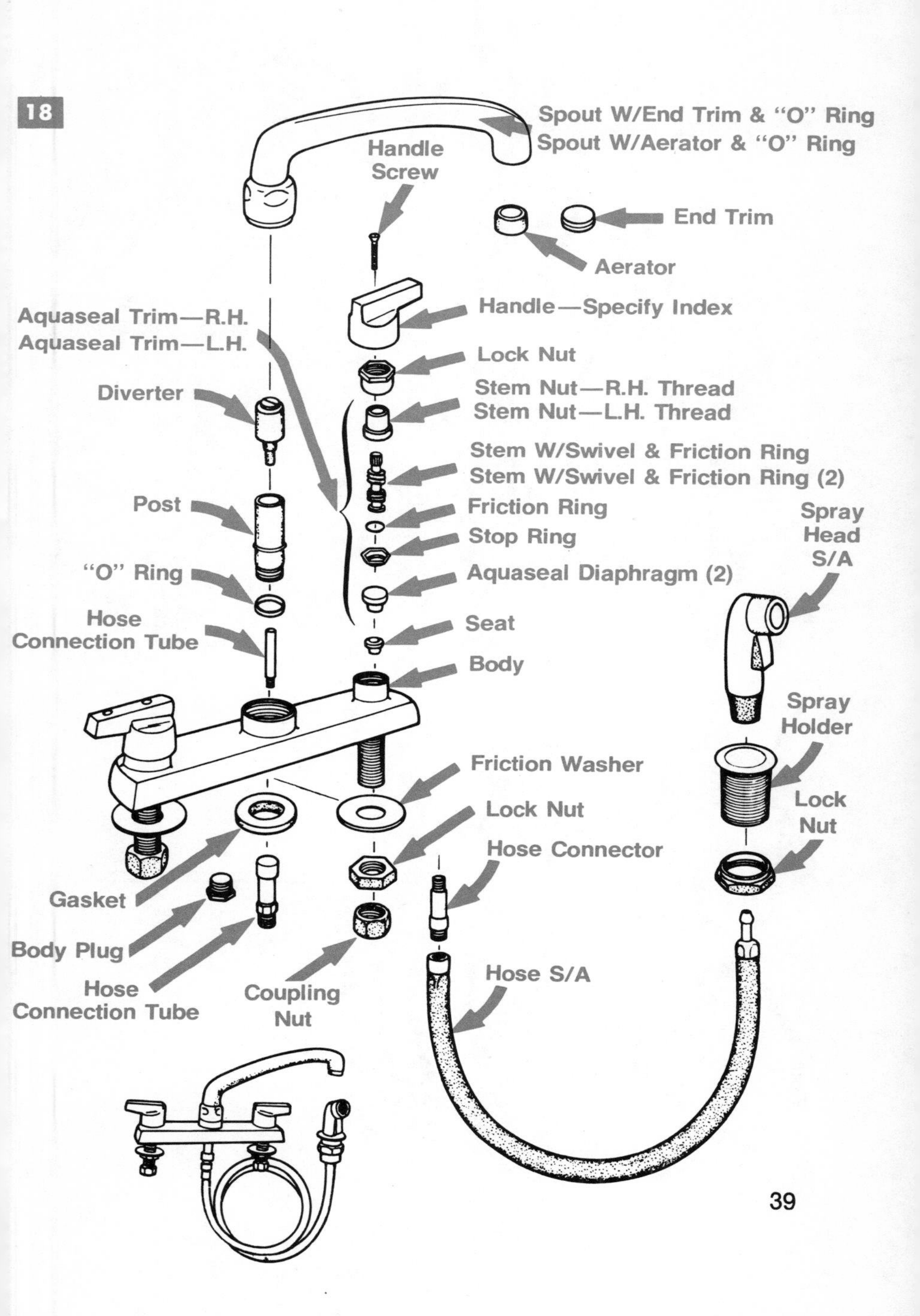
18
Spout W/End Trim & "O" Ring
Spout W/Aerator & "O" Ring
Handle Screw
End Trim
Aerator
Handle—Specify Index
Aquaseal Trim—R.H.
Aquaseal Trim—L.H.
Lock Nut
Stem Nut—R.H. Thread
Stem Nut—L.H. Thread
Diverter
Stem W/Swivel & Friction Ring
Stem W/Swivel & Friction Ring (2)
Post
Friction Ring
Stop Ring
Spray Head S/A
"O" Ring
Aquaseal Diaphragm (2)
Hose Connection Tube
Seat
Body
Spray Holder
Friction Washer
Lock Nut
Lock Nut
Hose Connector
Gasket
Body Plug
Hose Connection Tube
Coupling Nut
Hose S/A

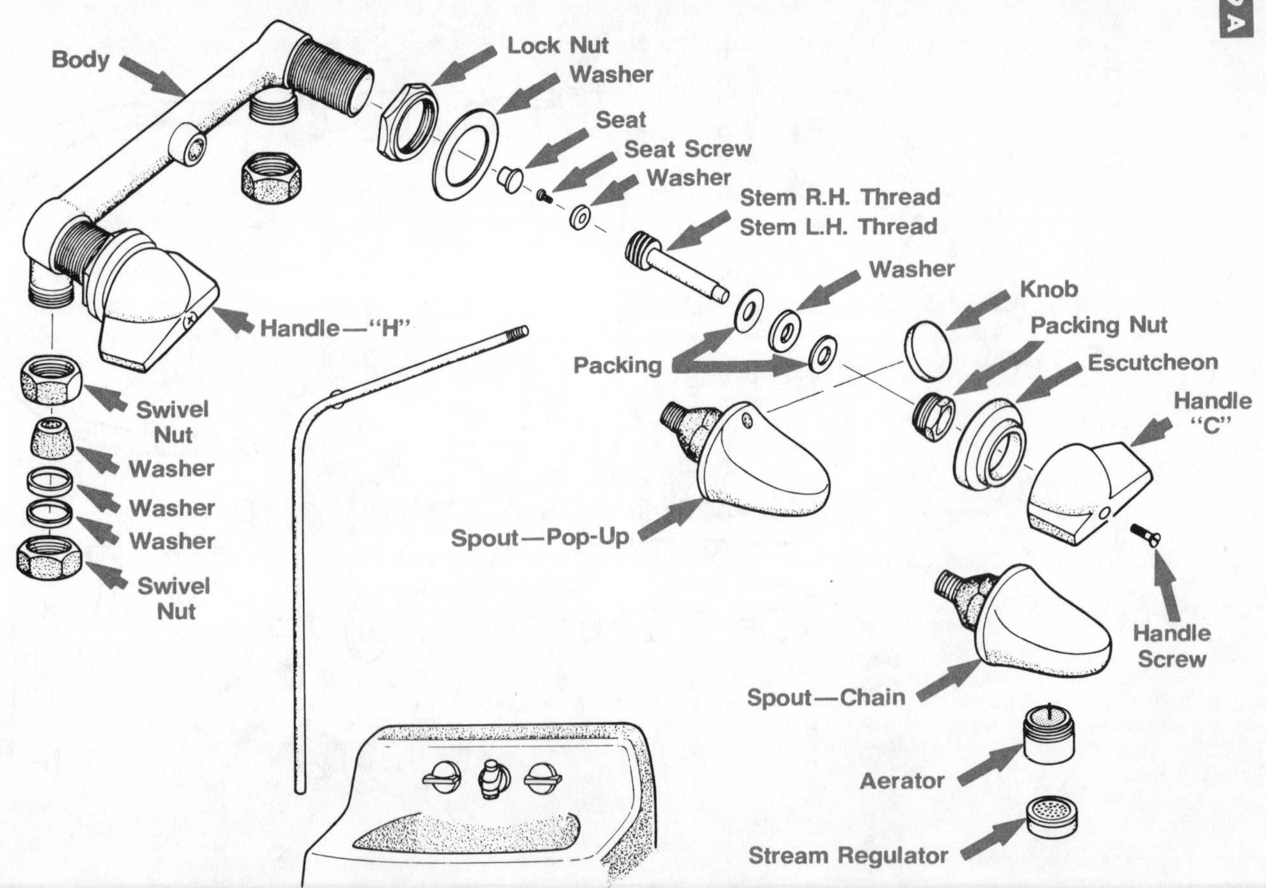
Body
Lock Nut
Washer
Seat
Seat Screw
Washer
Stem R.H. Thread
Stem L.H. Thread
Washer
Knob
Packing Nut
Escutcheon
Handle "C"
Handle—"H"
Packing
Swivel Nut
Washer
Washer
Washer
Swivel Nut
Spout—Pop-Up
Handle Screw
Spout—Chain
Aerator
Stream Regulator

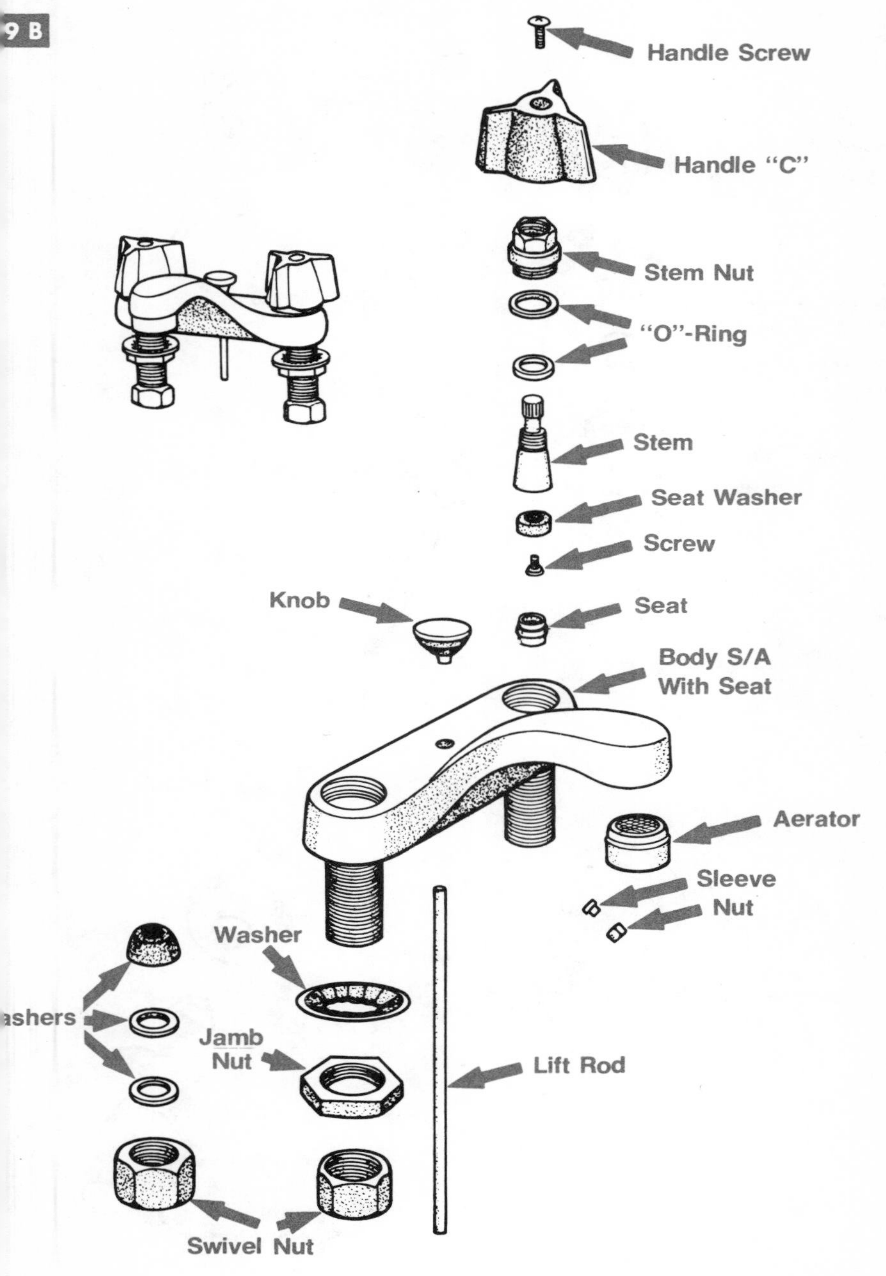
Handle Screw
Handle "C"
Stem Nut
"O"-Ring
Stem
Seat Washer
Screw
Knob
Seat
Body S/A
With Seat
Aerator
Sleeve
Nut
Washer
ashers
Jamb
Nut
Lift Rod
Swivel Nut

20

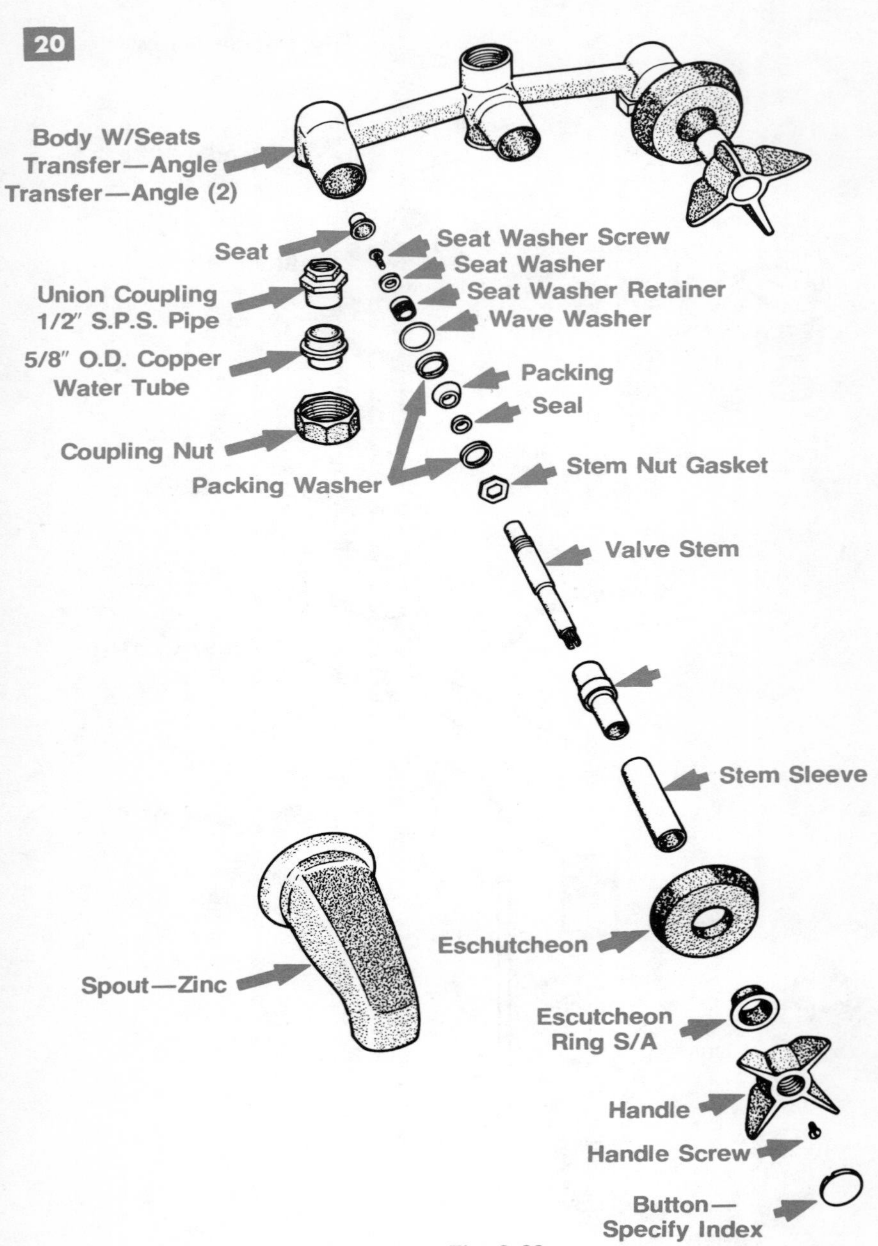
Body W/Seats
Transfer—Angle
Transfer—Angle (2)
Seat
Seat Washer Screw
Seat Washer
Seat Washer Retainer
Wave Washer
Union Coupling
1/2″ S.P.S. Pipe
5/8″ O.D. Copper
Water Tube
Packing
Seal
Coupling Nut
Stem Nut Gasket
Packing Washer
Valve Stem
Stem Sleeve
Eschutcheon
Spout—Zinc
Escutcheon
Ring S/A
Handle
Handle Screw
Button—
Specify Index

Fig. 3-20.

Single-Lever and Dripless Faucets

These faucets combine the convenience of a mixing faucet with a single-lever control. These single-lever and dripless faucets are very durable and seldom need repair when compared to compression faucets. Should repairs be necessary, usually simple replacements of O-rings, simple rubber valves, or replacement of a control assembly is all that is needed. Study Figs. 21, 22, 23, and 24 and become

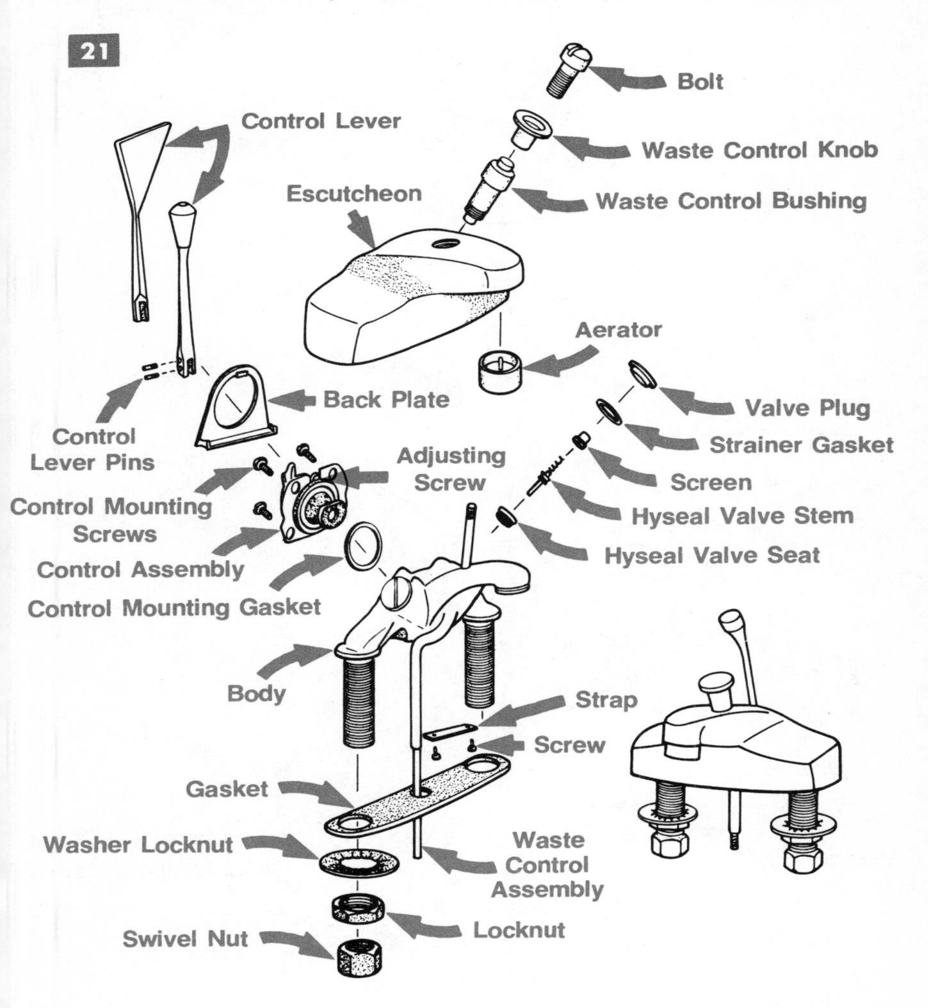

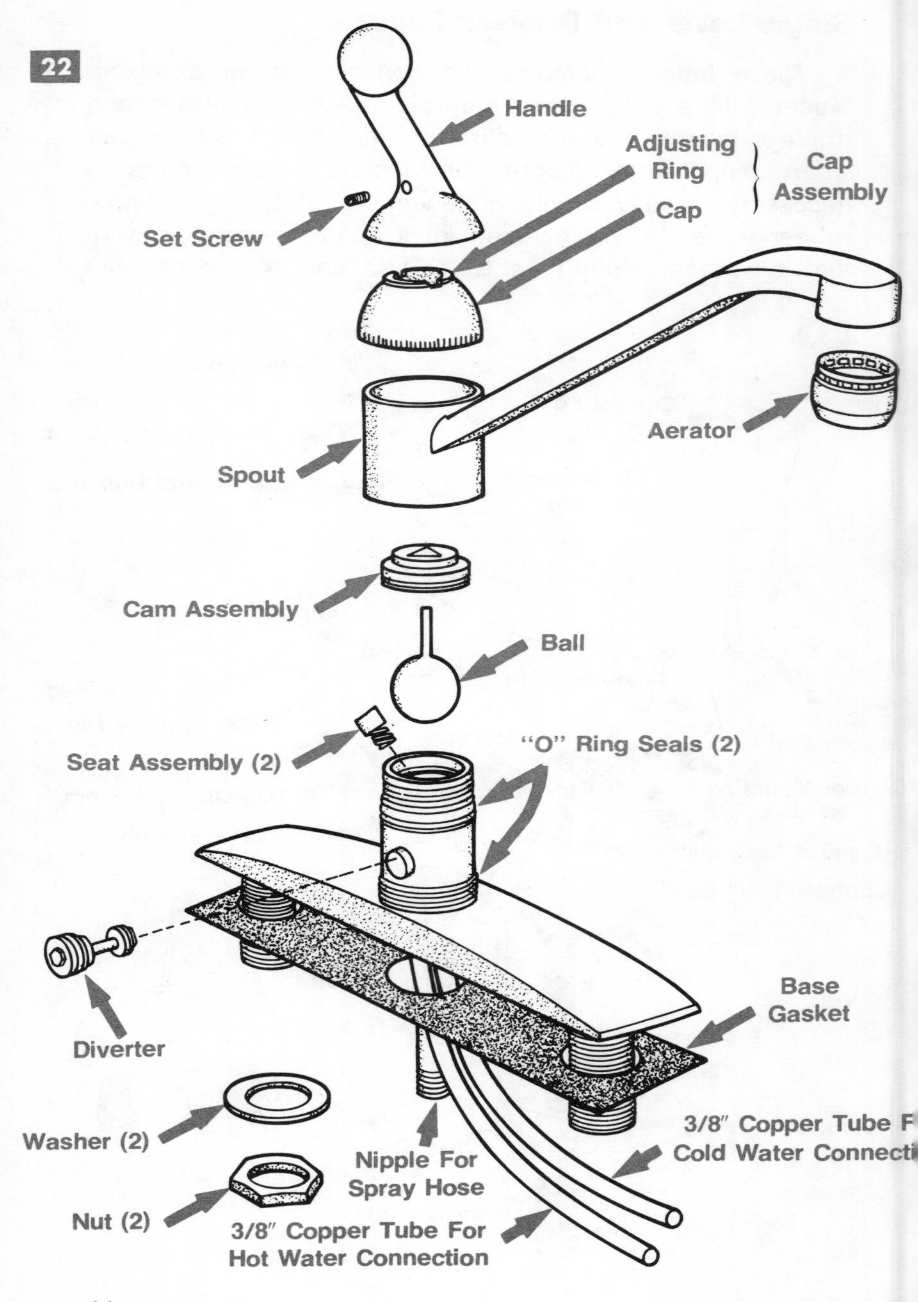
22
Handle
Adjusting Ring
Cap
Cap Assembly
Set Screw
Aerator
Spout
Cam Assembly
Ball
Seat Assembly (2)
"O" Ring Seals (2)
Base Gasket
Diverter
Washer (2)
Nipple For Spray Hose
3/8" Copper Tube F
Cold Water Connecti
Nut (2)
3/8" Copper Tube For Hot Water Connection

23

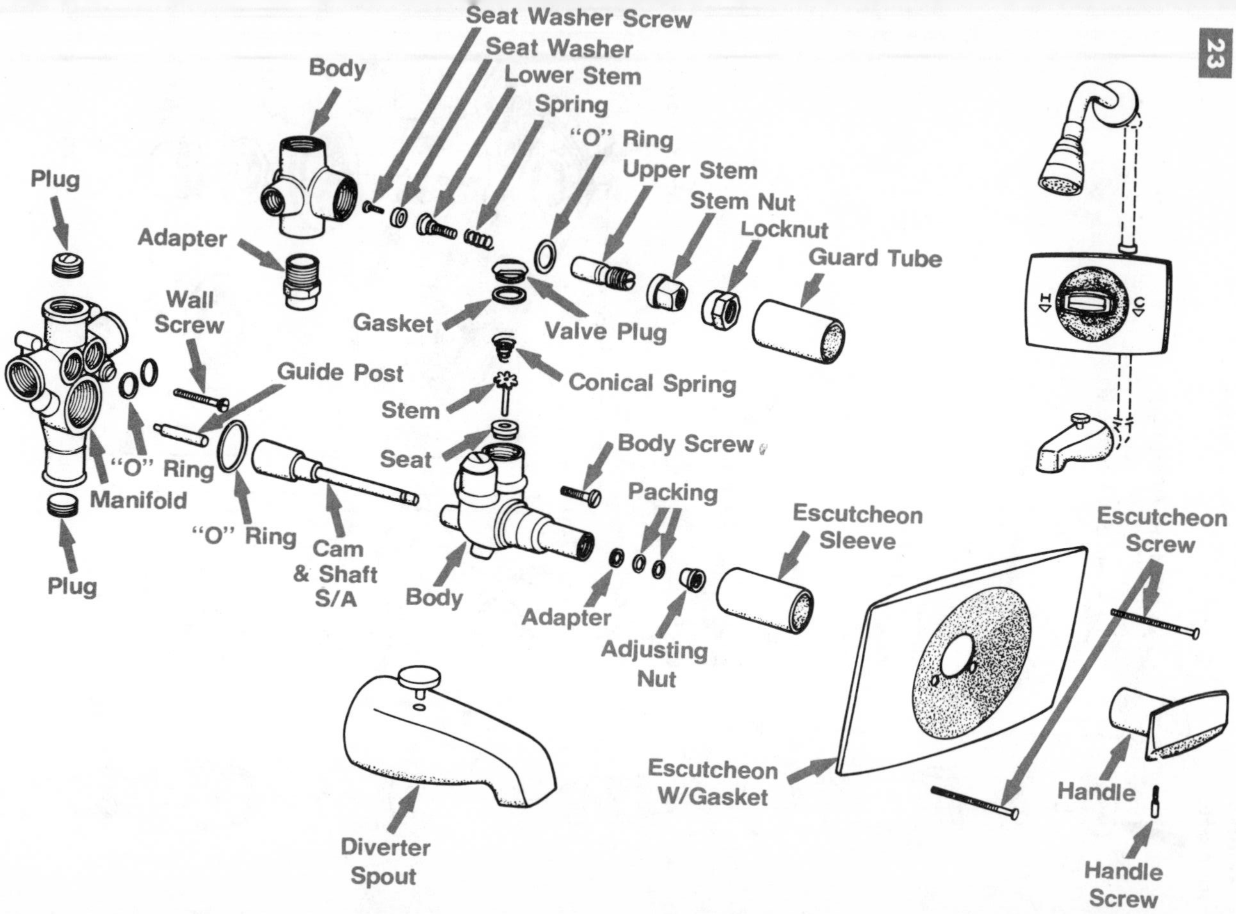

Seat Washer Screw
Seat Washer
Lower Stem
Spring
"O" Ring
Upper Stem
Stem Nut
Locknut
Guard Tube
Body
Plug
Adapter
Wall Screw
Gasket
Valve Plug
Conical Spring
Guide Post
Stem
Seat
Body Screw
"O" Ring
Manifold
"O" Ring
Cam & Shaft S/A
Body
Packing
Adapter
Adjusting Nut
Escutcheon Sleeve
Escutcheon Screw
Plug
Diverter Spout
Escutcheon W/Gasket
Handle
Handle Screw

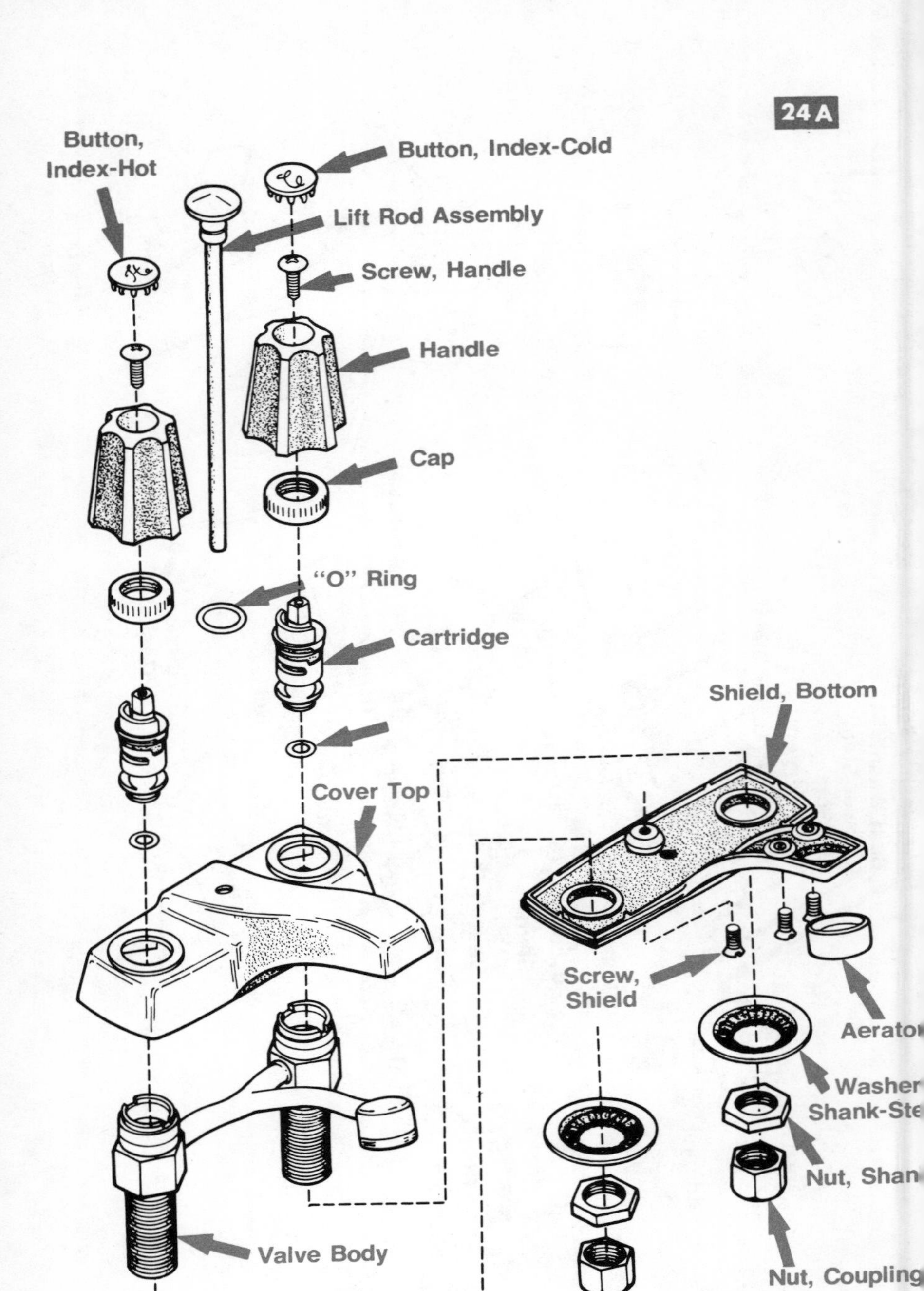
24 A
Button, Index-Hot
Button, Index-Cold
Lift Rod Assembly
Screw, Handle
Handle
Cap
"O" Ring
Cartridge
Shield, Bottom
Cover Top
Screw, Shield
Aerato
Washer
Shank-Ste
Nut, Shan
Valve Body
Nut, Coupling

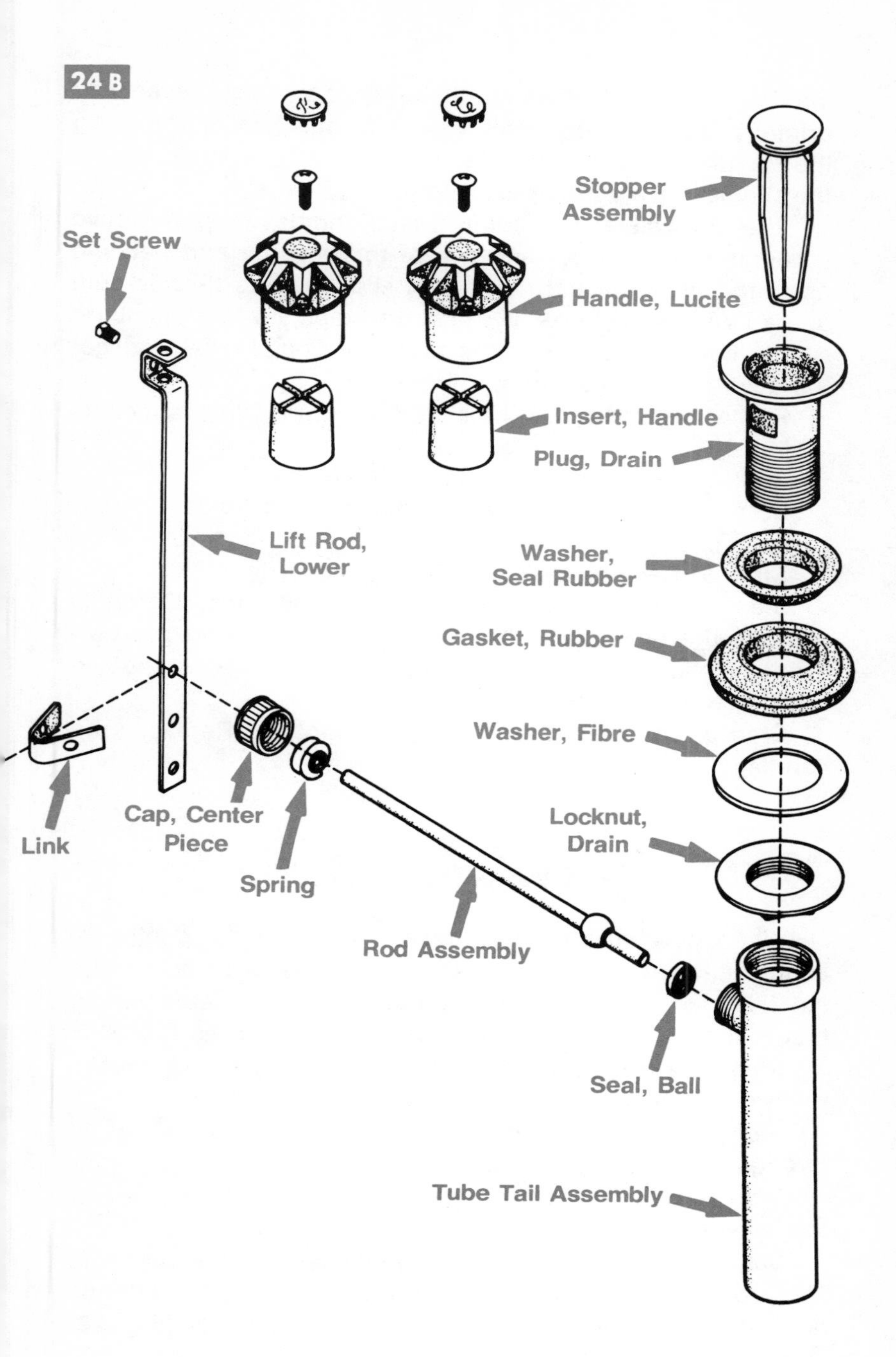
24 B
Set Screw
Stopper
Assembly
Handle, Lucite
Insert, Handle
Plug, Drain
Lift Rod,
Lower
Washer,
Seal Rubber
Gasket, Rubber
Washer, Fibre
Link
Cap, Center
Piece
Spring
Locknut,
Drain
Rod Assembly
Seal, Ball
Tube Tail Assembly

familiar with each part of these typical single-lever and dripless faucets. Disassembly usually consists of removing the handle, then various plates, ɹps, or spouts to get at the sliding-valve parts or assembly.

The dripless faucet found in many homes does not have a compression washer and a seat to wear out and become damaged. It has a cartridge with sliding parts that seldom go bad and cause leaks. If a leak develops, go to a plumbing supply house that carries the brand of your faucet and get replacement O-rings and cartridges (Fig. 17).

Replacing cartridges in several different dripless models is similar:

Step 1. Remove the handle screw and handle.

Step 2. Loosen and remove the cap that holds the cartridge in place.

Step 3. Remove the old cartridge and O-rings.

Step 4. Insert a new cartridge and O-rings. Carefully observe the placement and fit of the O-rings and the cartridge. Make sure everything fits just the same as the old one that was removed.

Step 5. Replace the cap and tighten. Replace the handles.

FAUCET REPLACEMENT

When purchasing a new faucet, be sure the center-to-center measurements of the shanks are the same as for the old faucet (Fig. 25). Once you have bought a replacement faucet that will fit your sink, it can be installed as follows:

Step 1. Turn off water supply. Open faucet to relieve pressure and to drain water from line.

Step 2. Disconnect and remove old faucet and clean all old putty from sink (Fig. 26).

Step 3. Set faucet into holes in sink and fasten faucet to sink with steel shank washers and shank nuts (Fig. 27).

Step 4. Connect the water-supply lines to the faucet with "hot" on left side and "cold" on right side. NOTE: Fittings and adapters to make the necessary connections (Fig. 28)

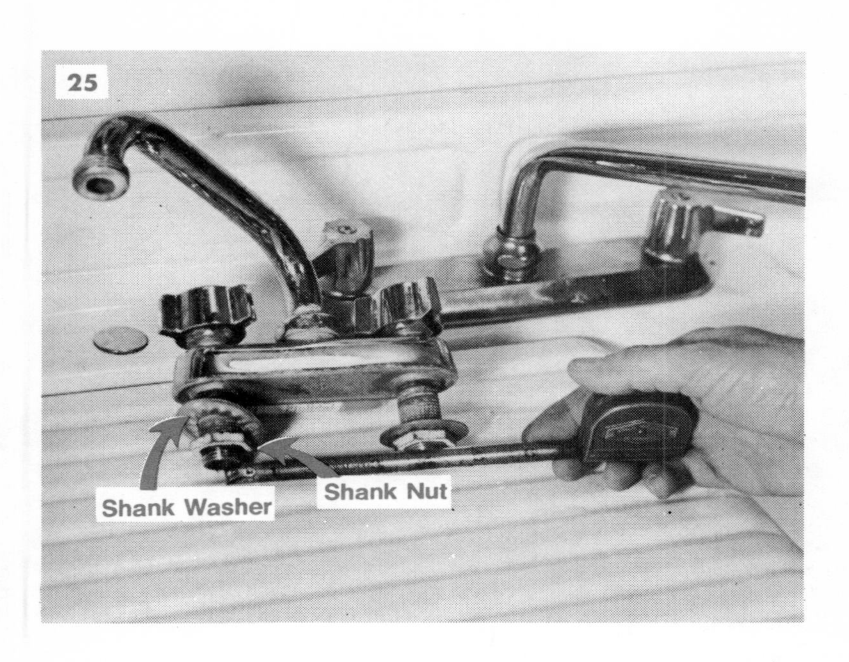
25
Shank Washer
Shank Nut

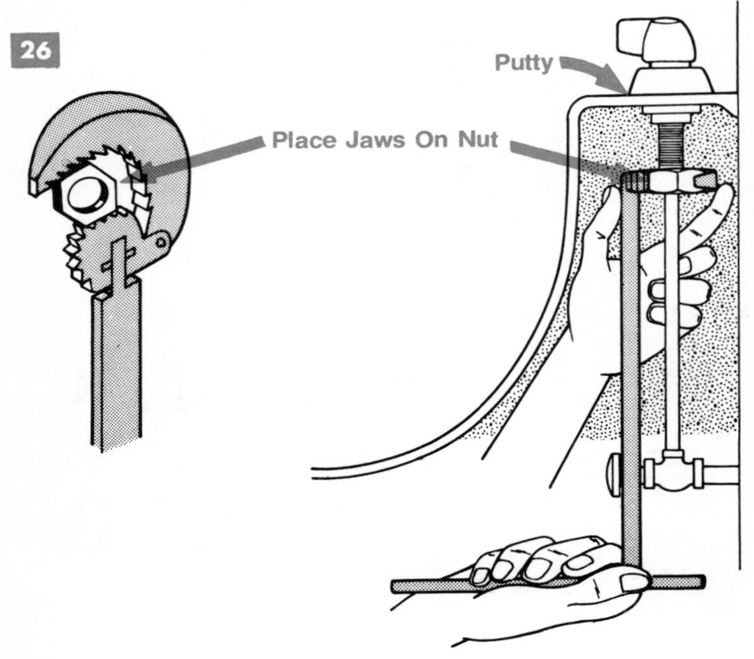
26
Putty
Place Jaws On Nut

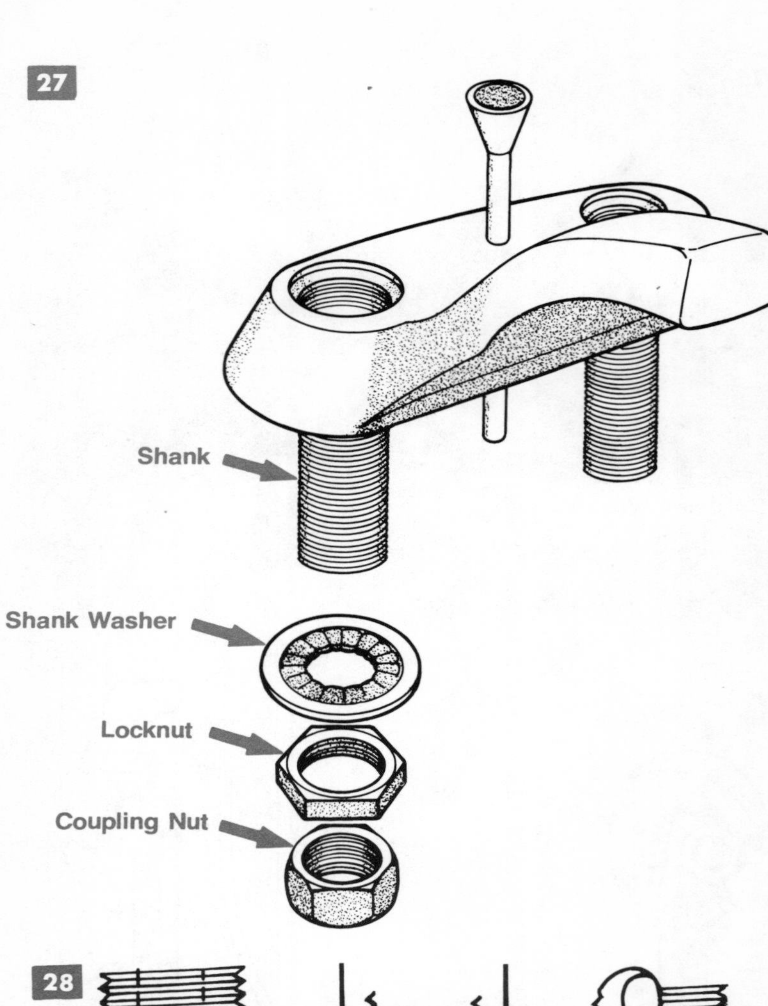
27
Shank
Shank Washer
Locknut
Coupling Nut

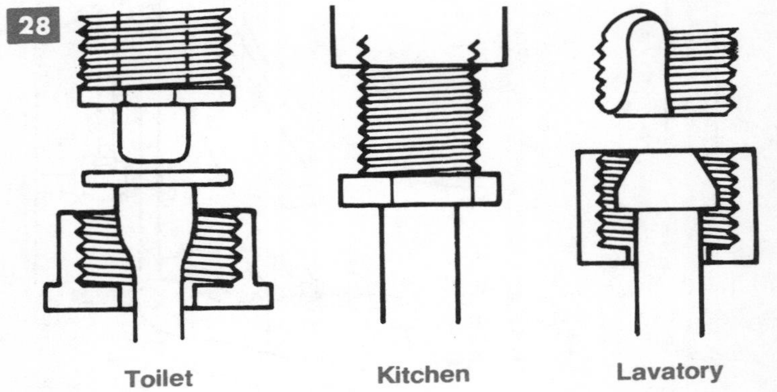
28
Toilet
Kitchen
Lavatory

are usually available at the store where you bought your faucet. In an earlier figure, it showed a small compression ring called a ferrule in the shut-off valve that makes a watertight seal around the supply pipe. This goes to a faucet shank. Whenever loosened, ferrules should be replaced. When doing so, it is a normal tendency to tighten the compression nut too tight thus crushing the pipe with the ferrule. To properly install the ferrule, tighten the locknut finger-tight. Then with an adjustable wrench tighten one-half turn. Turn on water to test. If it seeps or leaks, tighten a quarter turn at a time until it stops.

Step 5. Remove stopper assembly and drain plug from tube tail assembly. Remove drain locknut fiber washer and rubber gasket from drain plug. This is to be used at top of drain opening, under the flange of the drain plug, to prevent leakage of water between drain plug and drain hole (Fig. 24).

Step 6. Apply a small amount of plumbers putty around drain hole in sink. Assemble drain plug with rubber seal washer through drain hole of sink. Attach rubber gasket with round portion up, the fiber washer, and the drain locknut. Do not tighten the drain locknut.

Step 7. Assemble tube tail assembly to drain plug. Use thread sealing compound and tighten. Position assembly so that the side boss is pointed toward the rear of the sink. Tighten drain locknut to secure entire assembly to the sink.

Step 8. Make sure that ball seal is positioned in side boss of tube tail assembly and attach rod assembly. Tighten center piece cap (Fig. 24).

Step 9. Insert stopper assembly into drain plug.

Step 10. Remove link halfway from rod assembly. Attach the lower lift rod between the arms of the link and reassemble the link to the rod assembly. Insert the lift rod assembly through the pop-up hole in top of the faucet and through holes in the top of the lower lift rod. Tighten the set screw against the upper lift rod. Adjustment for height of knob above the faucet, and the center distance between the pop-up rod and the drain plug may now be made (Fig. 24).

Step 11. Turn on water supply and check for leaks. If leaks occur, tighten connections.

Step 12. Flush faucet thoroughly to remove any foreign particles which may possibly be in the water lines or faucet. To do this, remove the aerator from the spout and run water through faucet. Replace the aerator.

SWING NECK SPOUTS

Swing neck spouts on mixing faucets sometimes develop leaks around the neck where the spout swings in the faucet casting or housing (Fig. 29). To repair, loosen the locknut and slip out the spout. Replace the O-rings, insert the spout, and tighten the locknut.

Many faucet repairs like bad compression washers or worn or cut seats can be avoided by carefully turning faucets off each time. Too much shutting-off force simply shortens the life of the washer by smashing and cutting it. Not com-

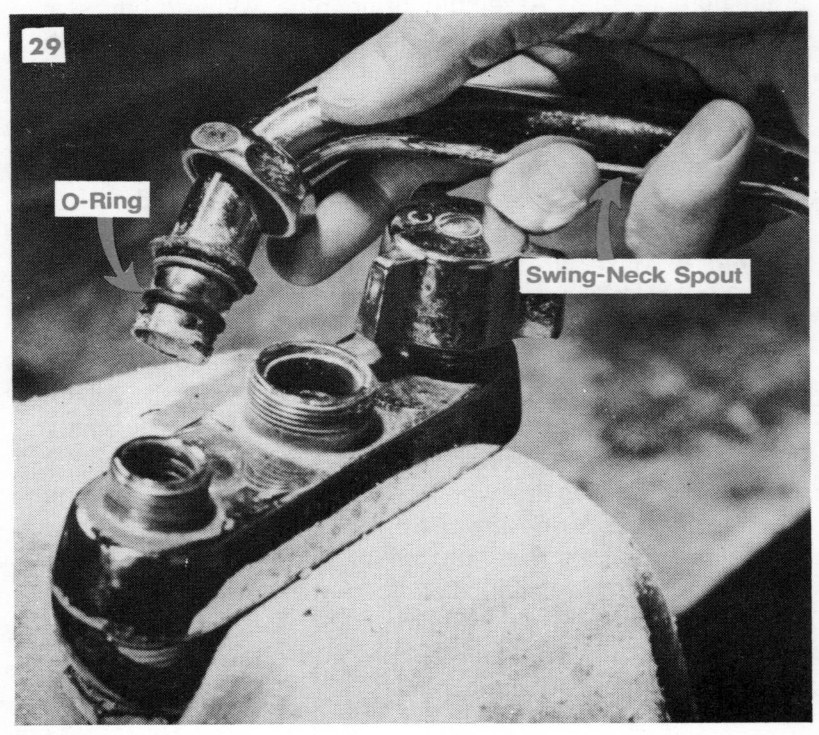

pletely shutting off a faucet is just as bad if not worse. Leaving a faucet drip will allow water to erode the washer and the chrome or brass seat. A stream of water squirting between a washer and a seat will literally cut a groove in the metal seat if left for several days or weeks. So treat your faucets with care; fix them as soon as any sign of a leak appears and they will give you long service.

30

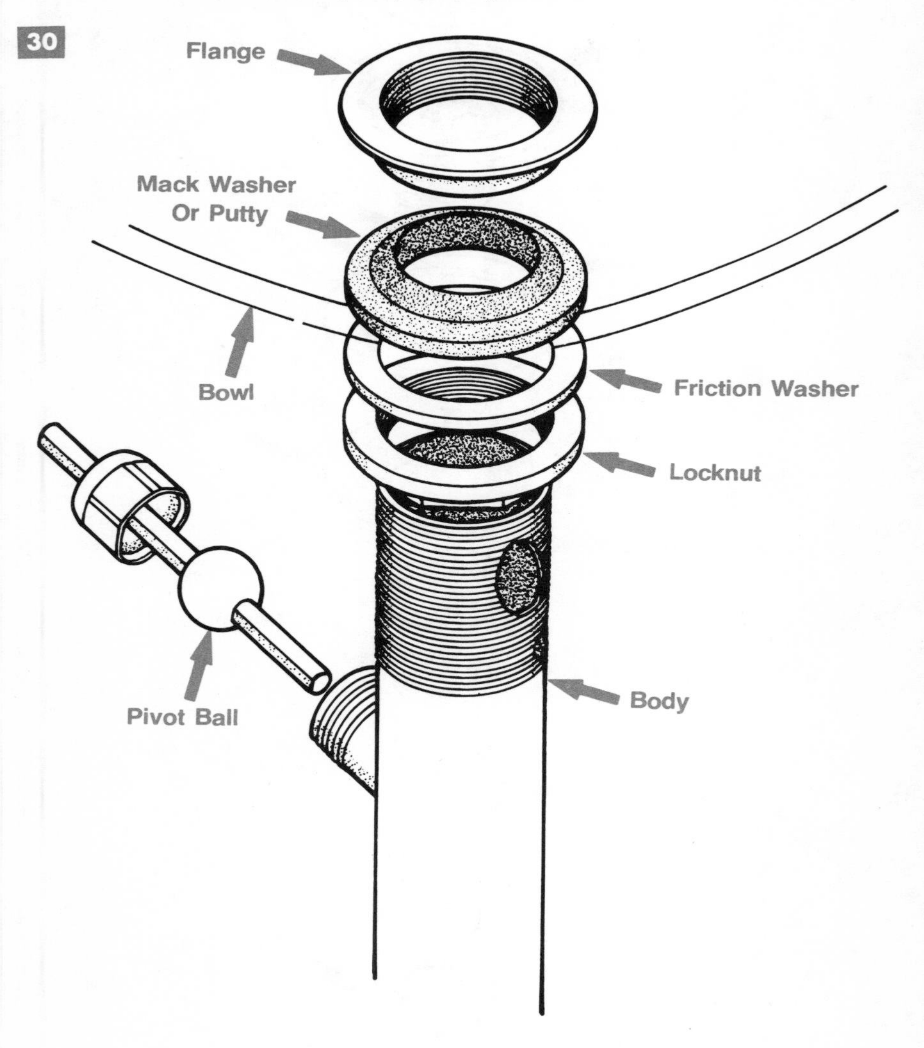

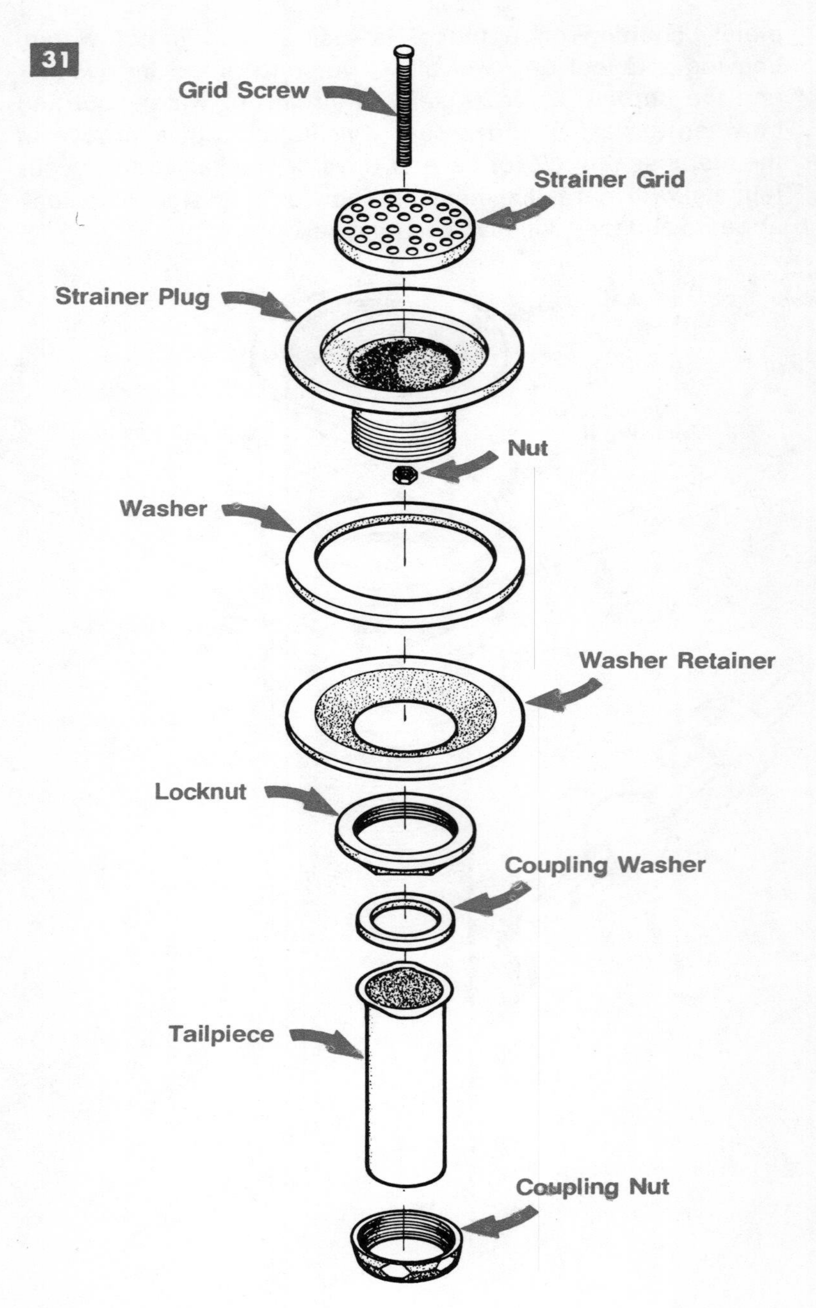
31
Grid Screw
Strainer Grid
Strainer Plug
Nut
Washer
Washer Retainer
Locknut
Coupling Washer
Tailpiece
Coupling Nut

DRAIN AND STOPPER PROBLEMS

A leak or malfunctioning stopper in the bottom of a lavatory is no serious problem. Study Figs. 30 and 31 to learn the proper parts of the typical pop-up stopper.

If the drain in the bowl leaks on the floor, proceed as follows:

Step 1. Loosen the locknut under the bowl.

Step 2. Remove the ball rod.

Step 3. Remove the P-trap.

Step 4. Raise flange up and remove it. Some drain fittings will be one piece and can be lifted right out.

Step 5. Remove and replace mack washer or bed flange with plumbers putty by placing a ring of putty around under the flange between it and the top of the bowl.

Step 6. Put friction washer back on and tighten locknut to put a squeeze by the flange, on the mack washer or putty. If putty is used, some will squeeze out around the flange. Simply scrape this up with a finger or cloth.

Tub drains can be serviced much the same way (Figs. 30 and 31). Two of the most common tub drains and stopper assemblies are shown in Fig. 32 A and B. When working on tub fixtures, be sure to place newspapers or an old rug in the bottom of the tub to prevent scratching the bottom. Access to the tub pipes and valves can be made through an access panel that is provided. Generally this access panel is in another adjoining room or closet (Fig. 33).

BATHROOM VENTILATION

Aside from the laundry room, the bathroom is exposed to probably more moisture than any other room in the house. Showers and tub baths release many pounds of water daily into the air. If the bathroom is particularly damp, the windows fog heavily or paint begins to mildew, then a bath ventilation fan should be installed. Small bathroom ventilation ceiling fans are inexpensive and can be installed in a couple of hours (Fig. 34). It should vent into the attic and not through the roof. Attic ventilation louvers and openings will more than adequately carry away the moisture.

32 A

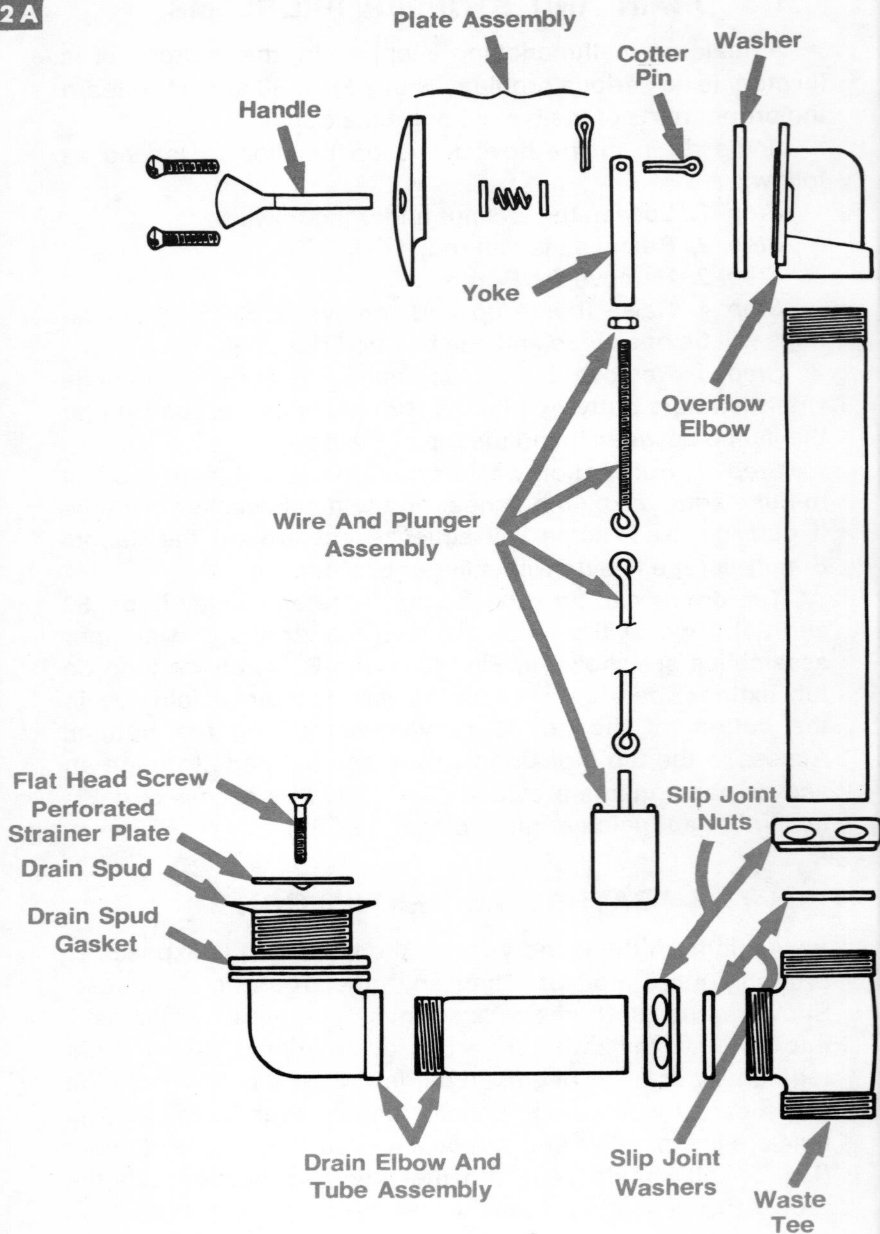
Plate Assembly
Washer
Cotter Pin
Handle
Yoke
Overflow Elbow
Wire And Plunger Assembly
Flat Head Screw
Perforated Strainer Plate
Drain Spud
Drain Spud Gasket
Slip Joint Nuts
Drain Elbow And Tube Assembly
Slip Joint Washers
Waste Tee

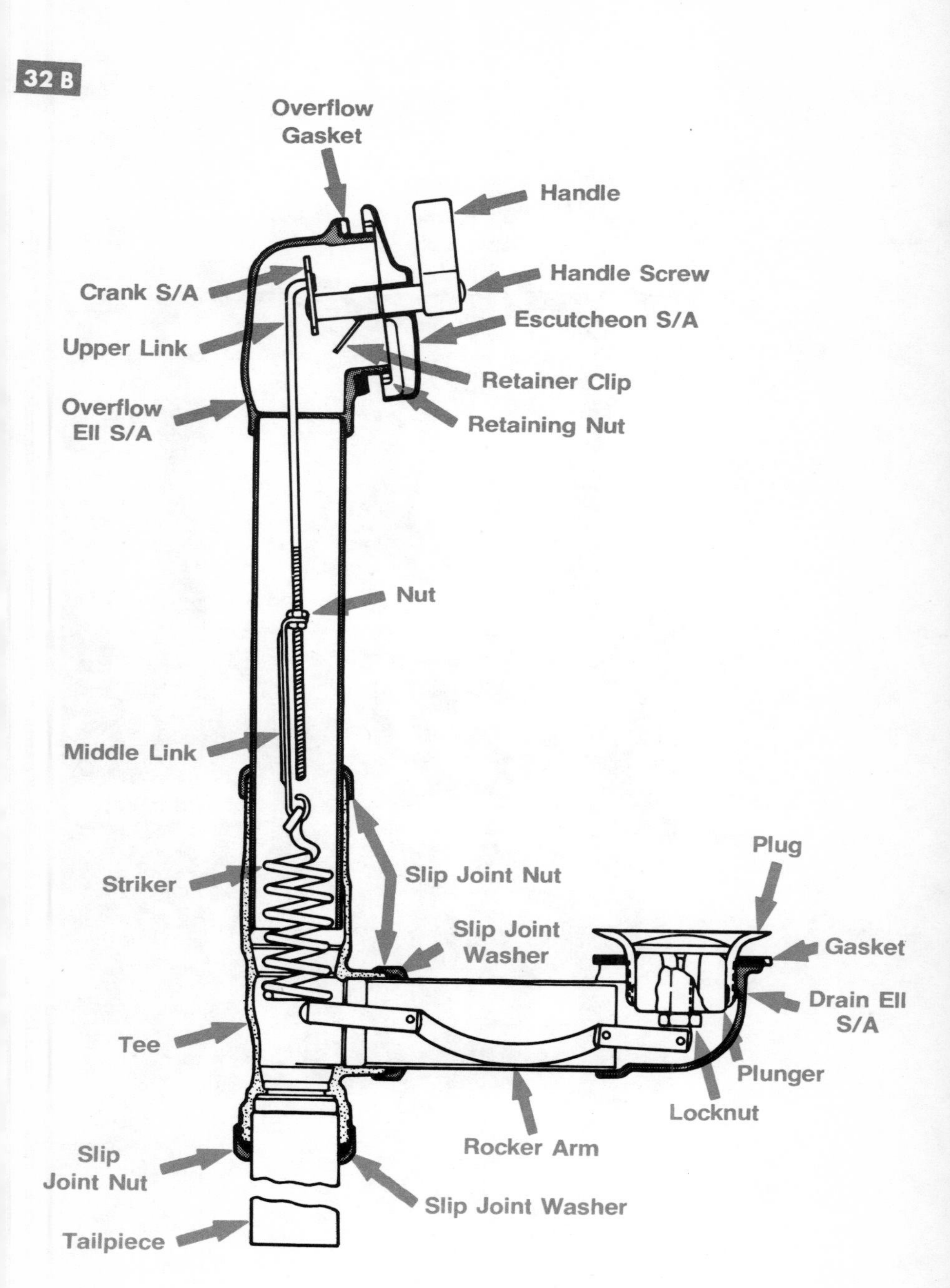
32 B
Overflow Gasket
Handle
Handle Screw
Crank S/A
Escutcheon S/A
Upper Link
Retainer Clip
Overflow Ell S/A
Retaining Nut
Nut
Middle Link
Plug
Slip Joint Nut
Striker
Slip Joint Washer
Gasket
Drain Ell S/A
Tee
Plunger
Locknut
Rocker Arm
Slip Joint Nut
Slip Joint Washer
Tailpiece

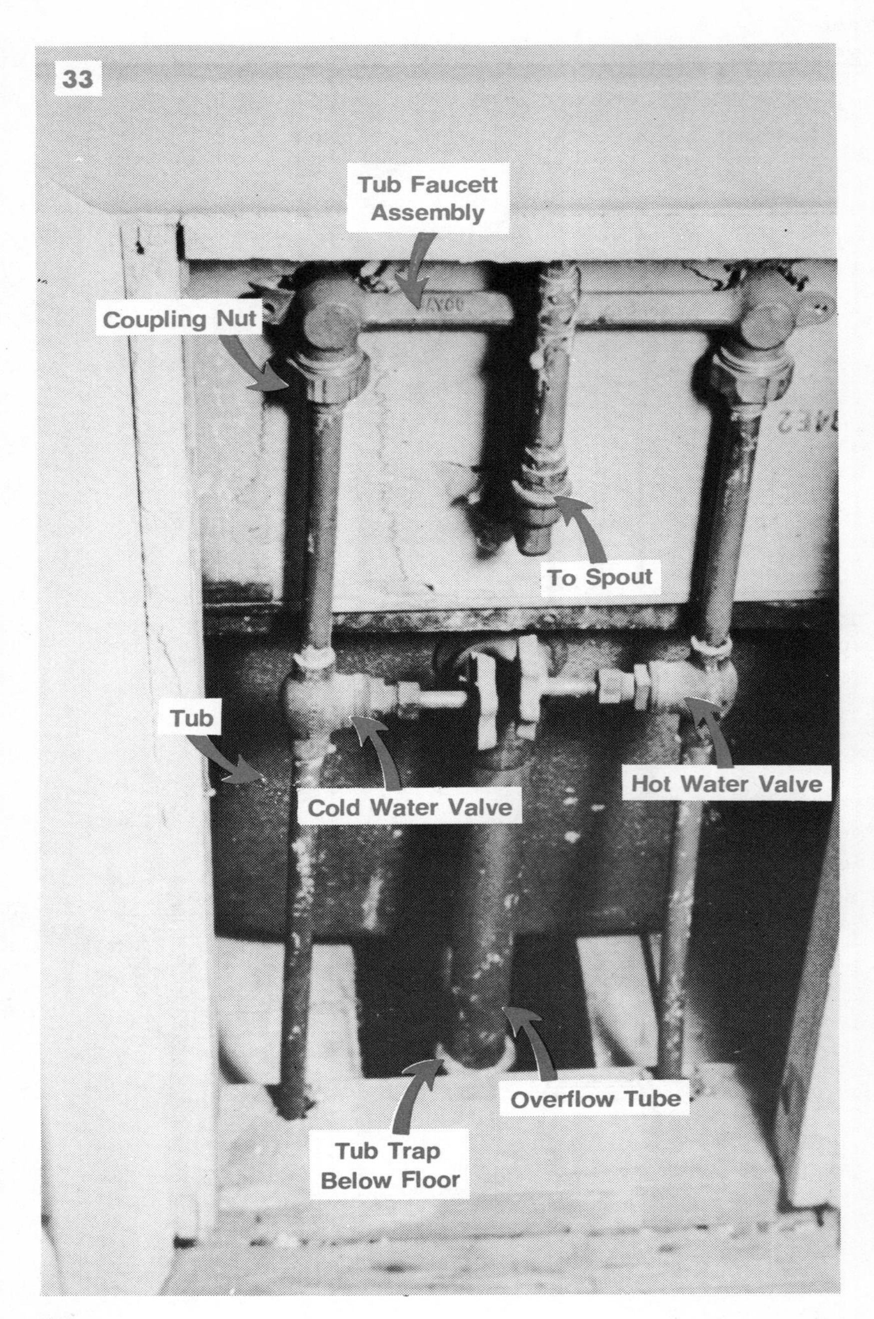
33
Tub Faucett
Assembly
Coupling Nut
To Spout
Tub
Cold Water Valve
Hot Water Valve
Overflow Tube
Tub Trap
Below Floor

VALVE AND FAUCET STEM LEAKS

If valves or faucets develop a leak around the stem, first try to tighten the packing nut one quarter turn clockwise with an end wrench (Fig. 35). If this fails, loosen and remove the packing nut. Dig out or remove all old packing or packing washer. Replace the packing with a packing washer just like the one that was removed or use a graphite-string type packing. Wrap enough of the rolled graphite packing or graphite string packing around the stem and into the bonnet of a valve or stuffing box of a faucet so that the packing nut will barely start with one thread catching. Tighten the packing nut until some drag or friction is felt when turning the faucet handle or valve handwheel.

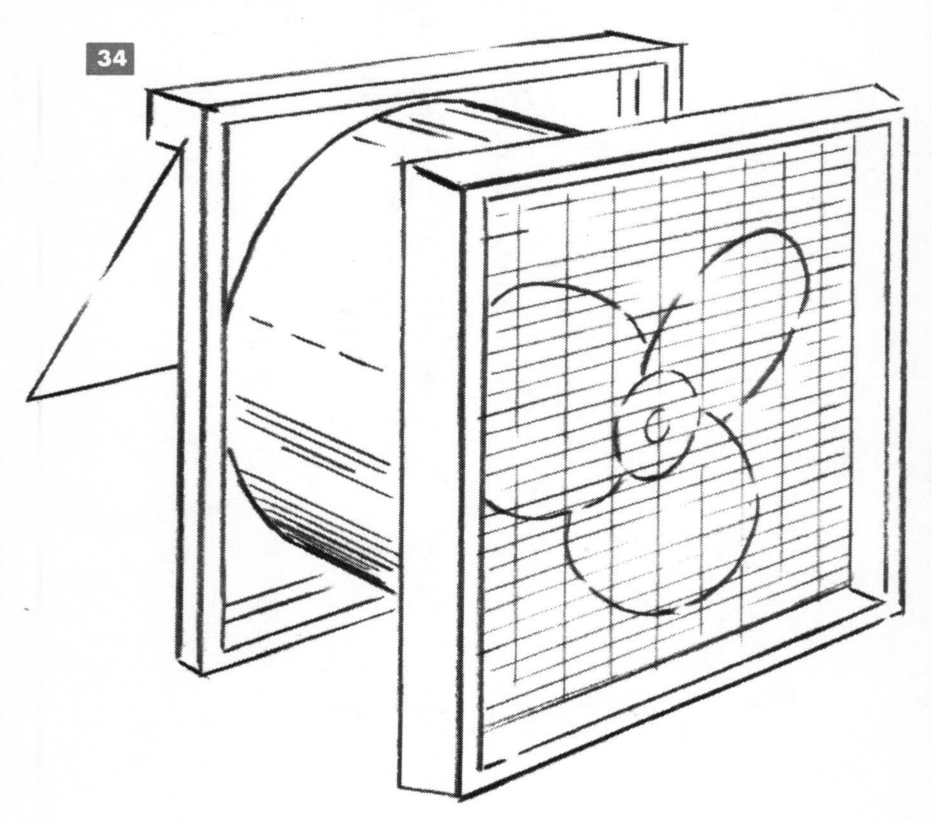

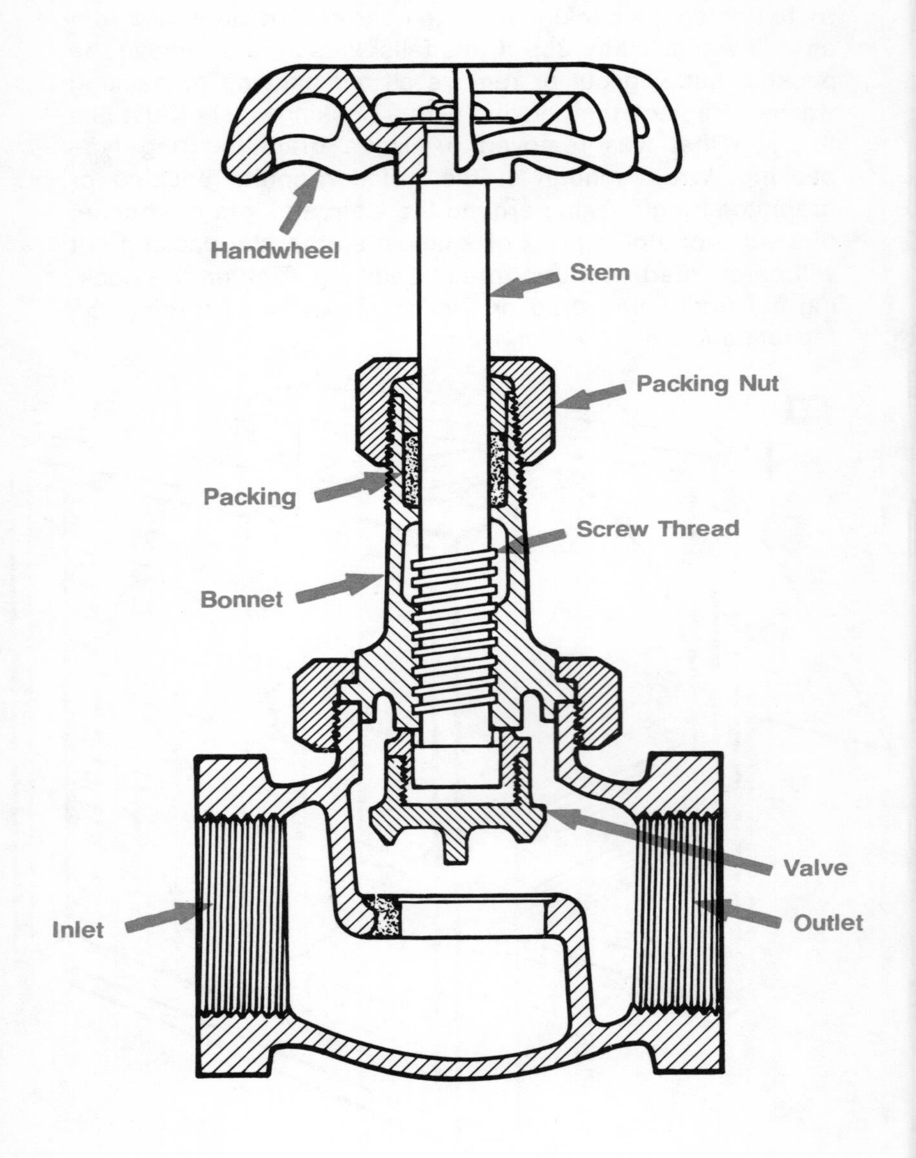
35
Handwheel
Stem
Packing Nut
Packing
Screw Thread
Bonnet
Valve
Inlet
Outlet

SECTION

Plumbing Supplies

Many persons who attempt to repair anything, if they are not familiar with the job, will disassemble or tear it apart and then determine what parts are needed. This is not uncommon for do-it-yourself plumbing repairs since most people know very little about plumbing tools, parts, equipment, and supplies. How do you go about learning the supplies, terminology, and people associated with plumbing? It isn't too difficult. Most journeyman plumbers (union ones at least) must go through a five-year apprenticeship training program. They virtually know it all. There is no reason for you to know that much but there are a few things you should be aware of and do.

Regardless of where you go, you must know what you need and be able to call it by a name that makes sense to a salesperson. Try to determine the name of the manufacturer who made what you need. Get a model name or style name or number. If nothing else, measure dimensions of important things like distances of hole centers, locations of holes, heights, etc. Make a sketch of it or even take a polaroid picture of it and take that along. The best method if possible is to take the actual defective part or fixture with you when you go after the plumbing supplies. Before you go, try to make a detailed list of everything needed. Even then several things are often forgotten.

The salesperson that you talk to may or may not be able to give much help. Of course, if you have worked with plumbing much you will be at ease in even the largest supply house and be able to make your needs known. If you are like everyone else you will want to ask a lot of questions and cling to every word of advice or suggestions that is mentioned by the salesperson or counterman.

First of all, regardless of where you go, lay it on the line and ask for help. Explain your problem as clearly and precisely as possible. Don't try to impress anybody with what you think or wish you knew.

The most help can be obtained by going to a plumbing supply house or local plumbing shop. Some large plumbing supply houses sell to mechanical contractors and plumbing shops and might not appreciate your small business. But try them anyway. Many of them genuinely want it. Some plumbing shops won't take the time with you either. However there are some shops that like to sell to the do-it-yourself plumbing trade. Try them all. A disadvantage of going to most plumbing shops is that normal business hours are often five days a week from 8:00 A.M. to 4:30 P.M. The average person can't get to them during those hours. That pretty well leaves you with the hardware and large chain department stores that have plumbing departments and who are open until 9:00 P.M.

Sometime when there is no need or emergency, stop in a good hardware store and slowly browse around the plumbing department (Fig. 1). Much can be learned by carefully studying what they have on display. Read labels, directions and instructions. Part of your need for plumbing knowledge is just knowing what supplies are available.

When you really need plumbing supplies seek a floor salesperson and ask for help. There is no guarantee that the person who helps you will know much more about plumbing than you. But between the two of you, something can always be worked out.

SECTION

Buying a used Home—Plumbing Considerations

Along with the many other reasons of inspecting a prospective used home, plumbing inspections on your part should almost come first. Simply because along with such problems as a leak in the roof or a leak in the basement, bad plumbing problems can be from troublesome to a nightmare.

First check the water for quality. Taste it; is it well water? What kind of conditioning equipment is in use? Find out the routine for servicing or recharging any conditioning equipment. Make sure any conditioning equipment is not removed from the house; it goes with it. Ask about the water supply throughout the year if a well is in use. Then visit several neighbors and inquire about their wells. Find out what type of pump is in use and when it was installed. Turn on a faucet and watch the pump work by the pressure gauge. Find out the cut-in and cut-out pressures. The gauge must work for this test. Have the owner locate the well for you.

If the house has city water, find out the monthly charges for water. Does this include a sewer-usage fee? If the house is on a well, where is the nearest city water line?

Flush each toilet several times and determine if it flushes quickly or sluggish. Look at the ballcocks in each tank to see if they operate properly. If there is a septic tank, find out when it was pumped out last. Make the owner carefully locate the septic tank and the layout of the finger system. Draw a sketch of this and put in measurements in feet.

Try all faucets to see if pressure is the same at each and also check each for leaks around the stem and washers. Leaks around the stem over a long period of time can be detected by a white substance around the stem and escutcheon. Notice if the drain at each fixture drains away quickly or sluggish, by filling the bowl and pulling the stopper.

Find out the age of the water heater, its size, and recovery time. If your family is two or more persons larger than the family living there, the water heater may be too small. Drain two or three gallons from the faucet at the base of the tank. Note its condition.

Check the location of all waste-line cleanout plugs. If you have to pull one out, how difficult would it be to reach it? Note their condition; do they show signs of seepage?

If the home has hot water heat, turn it on. Let it operate for at least an hour even in the middle of summer. Carefully check the boiler, pipes, and circulating pump for leaks. Find out if the circulating pump ever has had any kind of service and when. Get a detailed explanation of the operation and simple servicing of the system. Find out the operating pressures, etc.

If the house has gas or oil heat, insist on proof of total consumption and costs. With the price of fuel oil nowadays, only a fool would not find out the costs and useage if oil is used.

Check the sump pump if there is one. How old is it? What is its job? What is the condition of the screen? Does it discharge into a septic tank?

Inspect all pipes, fittings, and valves for leaks or corrosions caused by leaks.